WITHDRAWN

PERGAMON INTERNATIONAL LIBRARY
of Science, Technology, Engineering and Social Studies
The 1000-volume original paperback library in aid of education, industrial training and the enjoyment of leisure
Publisher: Robert Maxwell, M.C.

THE IRON BLAST FURNACE
Theory and Practice

THE PERGAMON TEXTBOOK
INSPECTION COPY SERVICE

An inspection copy of any book published in the Pergamon International Library will gladly be sent to academic staff without obligation for their consideration for course adoption or recommendation. Copies may be retained for a period of 60 days from receipt and returned if not suitable. When a particular title is adopted or recommended for adoption for class use and the recommendation results in a sale of 12 or more copies, the inspection copy may be retained with our compliments. The Publishers will be pleased to receive suggestions for revised editions and new titles to be published in this important International Library.

PERGAMON MATERIALS ADVISORY COMMITTEE

SIR MONTAGUE FINNISTON, PH.D., D.SC., F.R.S., *Chairman*
DR. GEORGE ARTHUR
PROFESSOR J. W. CHRISTIAN, M.A., D.PHIL., F.R.S.
PROFESSOR R. W. DOUGLAS, D.SC.
PROFESSOR MATS HILLERT, SC.D.
D. W. HOPKINS, M.SC.
PROFESSOR H. G. HOPKINS, D.SC.
PROFESSOR W. S. OWEN, D.ENG., PH.D.
PROFESSOR G. V. RAYNOR, M.A., D.PHIL., D.SC., F.R.S.
MR. A. POST, *Executive Member*

International Series on
MATERIALS SCIENCE AND TECHNOLOGY

Volume 31 *Editor*: D. W. Hopkins, M.Sc.

Other titles in the Series

BLAKELY: Introduction to the Properties of Crystal Surfaces

CHRISTIAN: The Theory of Transformation in Metals and Alloys, Part 1, 2nd Edition

COUDURIER, HOPKINS & WILKOMIRSKY: Fundamentals of Metallurgical Processes

DAVIES: Calculations in Furnace Technology

DOWSON & HIGGINSON: Elasto-hydrodynamic Lubrication, SI Edition

GABE: Principles of Metal Surface Treatment and Protection

GILCHRIST: Extraction Metallurgy, 2nd Edition

GRAY & MULLER: Engineering Calculations in Radiative Heat Transfer

HEARN: Mechanics of Materials

HULL: Introduction to Dislocations, 2nd Edition

KUBASCHEWSKI & ALCOCK: Metallurgical Thermochemistry, 5th Edition

MOORE: The Friction and Lubrication of Elastomers

MOORE: Principles and Applications of Tribology

PARKER: An Introduction to Chemical Metallurgy, 2nd Edition

PEARSON: A Handbook of Lattice Spacings and Structures of Metals and Alloys — Volume 2

RAYNOR: The Physical Metallurgy of Magnesium and its Alloys

REID: Deformation Geometry for Materials Scientists

SARKAR: Wear of Metals

SCULLY: Fundamentals of Corrosion, 2nd Edition

SMALLMAN & ASHBEE: Modern Metallography

UPADHYAYA & DUBE: Problems in Metallurgical Thermodynamics and Kinetics

WATERHOUSE: Fretting Corrosion

WILLS: Mineral Processing Technology

THE IRON BLAST FURNACE
Theory and Practice

by

J. G. PEACEY
Noranda Research Centre, Montreal, Canada

and

W. G. DAVENPORT
McGill University, Montreal, Canada

PERGAMON PRESS
OXFORD · NEW YORK · TORONTO · SYDNEY · PARIS · FRANKFURT

U.K.	Pergamon Press Ltd., Headington Hill Hall, Oxford OX3 0BW, England
U.S.A.	Pergamon Press Inc., Maxwell House, Fairview Park, Elmsford, New York 10523, U.S.A.
CANADA	Pergamon of Canada, Suite 104, 150 Consumers Road, Willowdale, Ontario M2J 1P9, Canada
AUSTRALIA	Pergamon Press (Aust.) Pty. Ltd., P.O. Box 544, Potts Point, N.S.W. 2011, Australia
FRANCE	Pergamon Press SARL, 24 rue des Ecoles, 75240 Paris, Cedex 05, France
FEDERAL REPUBLIC OF GERMANY	Pergamon Press GmbH, 6242 Kronberg-Taunus, Pferdstrasse 1, Federal Republic of Germany

Copyright © 1979 J. G. Peacey and W. G. Davenport

All Rights Reserved. No part of this publication may be reproduced, stored in a retrieval system or transmitted in any form or by any means: electronic, electrostatic, magnetic tape, mechanical, photocopying, recording or otherwise, without permission in writing from the publishers

First edition 1979

British Library Cataloguing in Publication Data

Peacey, J G
The iron blast furnace. – (Pergamon
international library: international series
on materials science and technology;
vol. 31).
1. Blast-furnaces
I. Title II. Davenport, W G
669'.1413   TN677   78-40823

ISBN 0-08-023218-3 (Hardcover)
ISBN 0-08-023258-2 (Flexicover)

*Printed and bound at William Clowes & Sons Limited
Beccles and London*

Contents

Preface   xi

Acknowledgements   xiii

1. A Brief Description of the Blast-Furnace Process   1

    1.1 Raw Materials   1
    1.2 Products   7
    1.3 Operation   9
    1.4 Improvements in Productivity   11
    1.5 Blast-furnace costs   12
    1.6 Summary   12
          Problems   14

2. A Look Inside the Furnace   16

    2.1 Behaviour in Front of the Tuyères   16
    2.2 Reactions in the Hearth, Tuyère Raceways and Bosh 18
    2.3 The Fusion Zone   19
    2.4 Reduction Above the Fusion Zone   21
    2.5 Kinetics of the Coke Gasification Reaction   23
    2.6 Reactions in Regions above the 1200 K Isotherm   23
    2.7 Reduction of Higher Oxides   23
    2.8 The Top Quarter of the Shaft and the Exit Gas   25
    2.9 Residence Times   25
    2.10 Burden Arrangements   26
    2.11 Summary   28
          Problems   29

3. Thermodynamics of the Blast-Furnace Process: Enthalpies and Equilibria 31

    3.1 Enthalpy Requirements in the Blast Furnace   31
    3.2 Critical Hearth Temperature   33
    3.3 Temperature Profiles in the Furnace: The Thermal Reserve Zone   35
    3.4 Free Energy Considerations in the Blast Furnace: The Approach to Equilibrium   36

Contents

3.5	Gas Composition Profiles in the Furnace: The Chemical Reserve Zone	38
3.6	Summary	40
	Problems	42

4. Blast-Furnace Stoichiometry — 44

4.1	The Stoichiometric Development	45
4.2	The Stoichiometric Equation	49
4.3	Calculations	50
4.4	Graphical Representation of the Stoichiometric Balance	51
4.5	Summary	56
	Problems	56

5. Development of a Model Framework: Simplified Blast-Furnace Enthalpy Balance — 58

5.1	Simplifications for an Initial Enthalpy Balance	58
5.2	The Enthalpy Balance	59
5.3	Heat Supply and Heat Demand	59
5.4	A General Enthalpy Framework	61
5.5	Summary	63
	Problems	63

6. The Model Framework: Combination of Stoichiometric and Enthalpy Equations — 65

6.1	Combining Stoichiometric and Enthalpy Equations: Calculations	66
6.2	Graphical Representation of the Combined Stoichiometric–Enthalpy Equation	68
6.3	A Graphical Calculation	70
6.4	Summary and Discussion of Stoichiometry/Enthalpy Graph	72
	Problems	74

7. Completion of the Stoichiometric Part of the Model: Conceptual Division of the Blast Furnace through the Chemical Reserve Zone — 75

7.1	The Blast Furnace as Two Separate Reactors	76
7.2	Stoichiometric Balances for the Bottom Segment	78
7.3	Stoichiometric Equation for the Wustite Reduction Zone	80
7.4	Discussion and Summary	81
	Problems	82

Contents

8. Enthalpy Balance for the Bottom Segment of the Furnace — 84

    8.1 Enthalpy Balance for the Bottom Segment — 84
    8.2 The Demand-Supply Form of the Enthalpy Equation — 86
    8.3 Numerical Development — 88
    8.4 Summary — 89
    Problems — 90

9. Combining Bottom Segment Stoichiometry and Enthalpy Equations: *a priori* Calculation of Operating Parameters — 91

    9.1 Example Calculations — 94
    9.2 Implications of the Equations — 96
    9.3 Graphical Representation of the Equations — 98
    9.4 A Graphical Calculation — 101
    9.5 Characteristics of the Operating Line — 104
    9.6 Summary — 106
    Problems — 106

10. Testing of the Mathematical Model and a Discussion of its Premises — 108

    10.1 Testing for Thermal Validity — 108
    10.2 Top-gas Temperature Calculation — 110
    10.3 Testing for Stoichiometric Validity — 113
    10.4 Testing for Thermodynamic Validity — 114
    10.5 Validity of the Model Assumptions and Predictions — 114
    10.6 Non-attainment of Equilibrium in the Chemical Reserve Zone — 117
    10.7 Thermal Reserve Temperature Effects — 117
    10.8 Summary — 120
    Problems — 121

11. The Effects of Tuyère Injectants on Blast-Furnace Operations — 123

    11.1 A General Injectant — 124
    11.2 Representing Injected Materials in the Overall Stoichiometric Equation — 126
    11.3 Representing Injected Materials in the Bottom Segment Stoichiometric Equation — 128
    11.4 Representing Injected Materials in the Bottom Segment Enthalpy Equation — 130
    11.5 A Form Convenient for Calculations — 133
    11.6 Example Calculations: I. Oxygen Enrichment — 133
    11.7 Example Calculations: II. Hydrocarbon Injection — 140
    11.8 Graphical Calculations (General Case) — 144
    11.9 Top-gas Composition with Hydrogen Injection — 148
    11.10 Discussion of Injection Calculations and Summary — 149
    Problems — 150

Contents

12. Addition of Details into the Operating Equations: Heat Losses; Reduction of Si and Mn; Dissolution of Carbon; Formation of Slag; Decomposition of Carbonates — 153

- 12.1 Stoichiometric Effects — 153
- 12.2 Enthalpy Effects — 157
- 12.3 Summary — 164
- Problems — 165

13. Summary of Blast-Furnace-Operating Equations: Comparison between Predictions and Practice — 167

- 13.1 Summary of Model Development Steps — 167
- 13.2 A Strategy for Computer Calculation — 173
- 13.3 Comparison of Model Predictions with Industrial Blast-furnace Data — 173
- 13.4 Effects of Blast Temperature, Tuyère Injectants, Metallized Ore and Metal Impurities on Coke and Blast Requirements — 176
- 13.5 Summary — 179

14. Blast-furnace Optimization by Linear Programming — 181

- 14.1 A Simplified Optimization Problem — 182
- 14.2 Graphical Representation of Cost Minimization — 184
- 14.3 Analytical Optimization Methods — 189
- 14.4 Computer Inputs and Outputs — 191
- 14.5 A More Complete Problem — 196
- 14.6 Summary — 202
- Problems — 203

Appendix I Tuyère Flame Temperature Calculations — 205

- AI.1 Flame Temperature Equations for Linear Programming — 208
- AI.2 Additional Items in the Calculations — 210

Appendix II Representing Complex Tuyère Injectants in the Operating Equations — 212

- AII.1 Gaseous Injectants with Known Heats of Combustion and Chemical Compositions — 212
- AII.2 Injectants with Known Weight Percentages of Carbon and Hydrogen and Known Heats of Combustion — 214

Appendix III Slag Heat Demands — 216

Appendix IV	Stoichiometric Data for Minerals and Compounds in Ironmaking	219
Appendix V	Enthalpies of Formation at Temperature T from Elements at Temperature T (H_T^f)	220
Appendix VI	Enthalpy Increment Equations for Elements and Compounds, $[H_T^\circ - H_{298}^\circ]$	222
Appendix VII	Numerical Values of E^B, Blast Enthalpy	224
Answers to Numerical Problems		225
List of Symbols		227
Index		231

Preface

The iron blast furnace is one of man's most useful tools. It provides the means by which iron is rapidly and efficiently reduced from ore and it is the basis for virtually all primary steelmaking. It is a significant item in the economy of any country.

The importance of the blast furnace has led it to be the object of many detailed studies. In spite of this, its complexity has prevented it from being fully understood. Reaction mechanisms and rates of individual steps (e.g. iron oxide reduction) have been established, but the blast-furnace process as a complete operation has defied quantitative description.

Several such descriptions have been attempted, the most successful being that developed by Rist and co-workers* based on mass and enthalpy balances and reaction equilibria. Unfortunately, the Rist explanations have not been as widely understood as might have been hoped. In addition, a number of empirical models based on industrial data have been devised for the process and these have found considerable importance in optimizing blast-furnace performance. In general, however, they are not instructive as to why the blast furnace behaves as it does.

In this text we have gathered existing theoretical, experimental and operational evidence about the blast furnace from which we have put together a mathematical description of its operation. Specifically, we have prepared a set of equations which (i) accurately describe stoichiometric and enthalpy balances for the process and which (ii) are consistent with observed temperatures and compositions in the furnace stack. Our equations have been devised on the basis of the Rist approach and they employ the ideas and experimental results of many other workers as are acknowledged in the reference sections.

---

*Rist's work appears in a series of papers in *Revue de Métallurgie* from February 1964 to April 1966, Volumes 61 to 63.

The equations are specifically designed to show the effects of altering any blast-furnace variable on the other operating requirements of the process. For example, they show analytically and graphically the effect which an increased blast temperature has on the enthalpy balance of the process and they indicate quantitatively how the coke and blast inputs to the furnace can be diminished to take advantage of this enthalpy effect.

Other important variables covered by the equations are hydrocarbon injection at the tuyères, oxygen enrichment of the blast, moisture, limestone decomposition, coke reactivity and metalloid reduction. The effects of many of these variables are illustrated numerically in the text while others are demonstrated in sets of problems which follow each chapter (numerical answers, page 225). The problems also serve to entrench each new concept as it arises in the text.

We hope that our approach will lead to an improved understanding of the blast furnace, and that readers of this text will better comprehend the separate and combined effects of each process variable. We also hope that our equations will be tested on operating blast furnaces and that their explanations and suggestions will be useful for optimizing furnace performances.

The text and the problems are used at McGill University for the first half of an iron-and-steelmaking course and they may be useful for this purpose elsewhere.

J. G. PEACEY
Noranda Research Centre

W. G. DAVENPORT
March 1979   McGill University

Acknowledgements

This work is largely based on ideas presented by Professor Andre Rist, Ecole Centrale des Arts et Manufactures, and we thank him for his inspiration and for his personal discussions of the subject. We would also like to express our sincere gratitude to the many metallurgists and engineers of the Iron and Steel Industry who have furnished us with up-to-date industrial data and who have, through plant visits and discussions, helped us to better understand industrial blast-furnace practice.

The encouragement of our colleagues at McGill University and at the Noranda Research Centre has been unflagging throughout the work. We would especially like to thank Dr. W. M. Williams, Dr. G. W. Smith, Dr. W. H. Gauvin and Dr. G. R. Kubanek in this regard.

The ideas in this text have developed during approximately 10 years of exposure to young metallurgy students. Our greatest gratitude goes to them for their enthusiasm, for their perseverance and most of all for their senses of humour during our many confrontations.

CHAPTER 1

A Brief Description of the Blast-Furnace Process*

The iron blast furnace is a tall, vertical shaft furnace which employs carbon, mainly in the form of coke to reduce iron from its oxide ores. The product is a liquid 'pig iron' (4–5% C, ½–1% Si) which is suitable for subsequent refining to steel. A schematic view of a typical blast-furnace plant is presented in Fig. 1.1 and a detailed blast-furnace profile is shown in Fig. 1.2. Operating details of large modern blast furnaces are given in Table 1.1.

The principal objective of the blast furnace is to produce molten iron of constant composition at a high rate. The only critical operating parameter is the temperature of the iron and slag which must be greater than 1700 K for these products to be tapped from the furnace in the molten state. Metal composition is not a critical feature of the blast-furnace process because virtually all blast-furnace iron is subsequently refined to steel. It is, however, controlled to steelmaking plant specifications by appropriate adjustments of slag composition and furnace temperature.

There are some 1000 blast furnaces in the world with a total productivity of approximately 500 million tonnes of molten pig iron per year.

1.1 Raw Materials

The raw materials of the blast furnace are (i) solids (ore, coke, flux) which are charged into the top of the furnace; and (ii) air which is blasted through tuyères near the bottom of the furnace. Hydrocarbon additives

---

*For detailed descriptions of blast furnace operations, the reader is directed to *Blast Furnace Ironmaking, 1977*, and *Blast Furnace Ironmaking, 1978*, W. K. Lu, Editor, McMaster University, Hamilton, Canada.

2 The Iron Blast Furnace

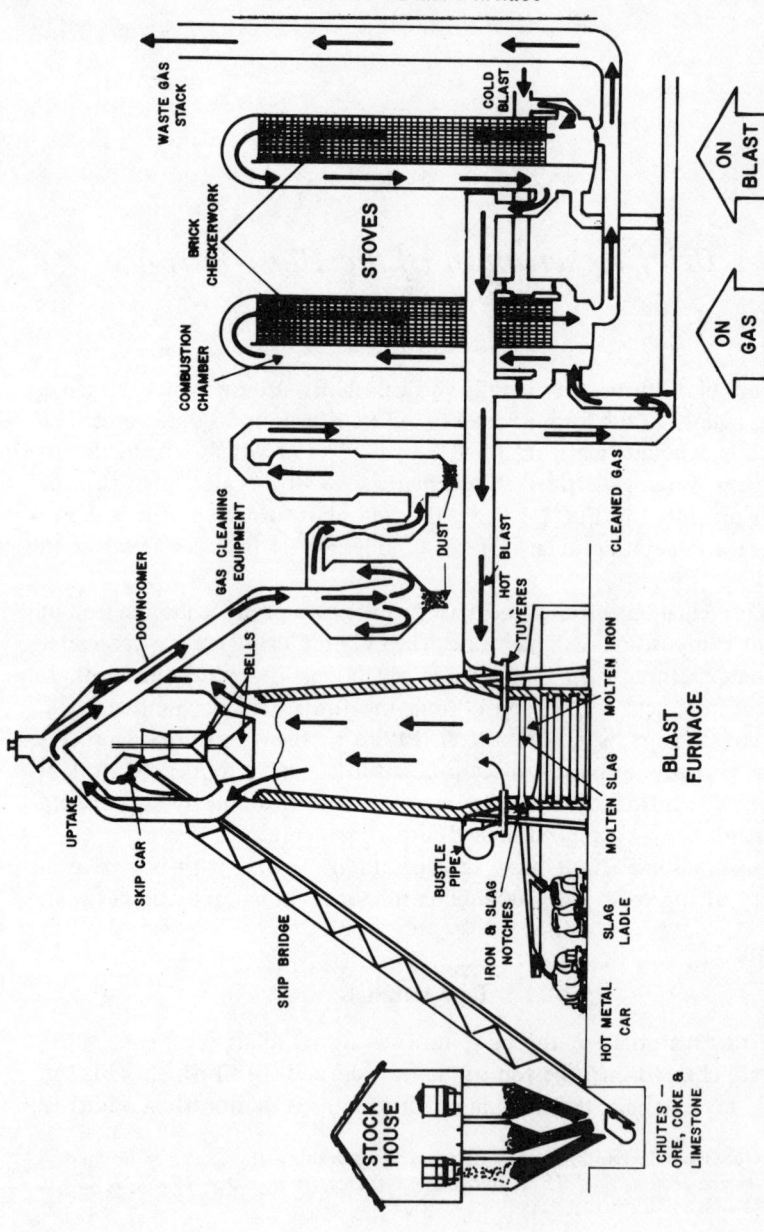

Fig. 1.1. Schematic cross-section of a blast-furnace plant showing materials-handling, charging, tapping, gas-handling and hot-blast equipment.

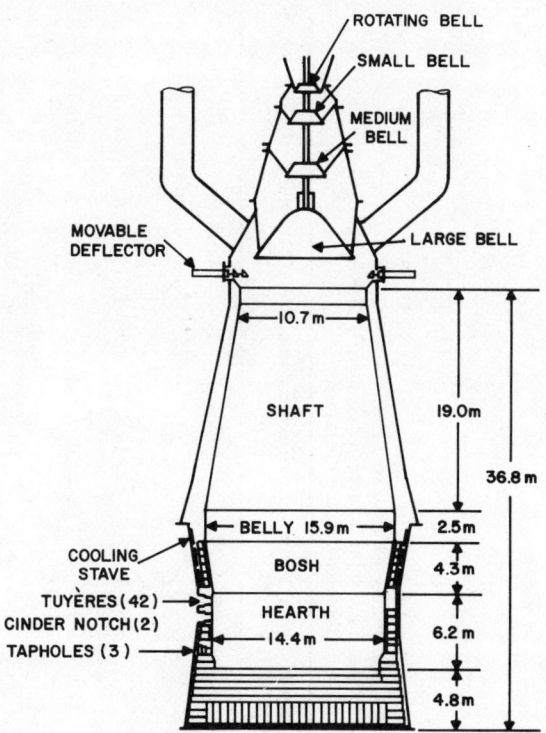

Fig. 1.2. Vertical profiles of Fukuyama number 5 blast furnace (Sugawara, 1976). Two features are notable: the quadruple bell system which indicates that the furnace is, like most modern units, being operated under pressure; and the movable deflectors around the throat which are used to obtain uniform burdening across the furnace. The working volume, tuyères to stockline, and the internal volume, hearth bricks to stockline, of this furnace are 3930 and 4620 m^3 respectively.

(gas, liquid or solid) and oxygen are also introduced through the tuyères. A representative materials balance is presented in Fig. 1.3.

The solid raw materials consist of:

(a) *Iron oxides*: usually hematite, Fe_2O_3, occasionally magnetite, Fe_3O_4. In modern operations the iron oxides are added in the form of (i) 1–2-cm diameter pellets produced from finely ground, beneficiated ore (5–10% $SiO_2 + Al_2O_3$, remainder Fe_2O_3);

TABLE 1.1
Operating Details of Modern Blast Furnaces.
The data for Fukuyama number 5 and Italsider number 5 are from Higuchi (1977) and De Marchi (1978). The remaining data were supplied directly by the blast-furnace operators.

Furnace identification		Burns Harbour 'D' 1977	Canadian 1977	Fukuyama no. 5 1976	Chiba no. 5 1976	Italsider no. 5 1976
Furnace production (pig iron)	tonnes day^{-1}	5700	2500	9900	4900	8100
Productivity (based on working volume)	tonnes day^{-1} m^{-3}	2.4	1.6	2.5	2.7	2.4
Dimensions of furnace						
Hearth diameter	m	10.7	8.5	14.4	11.1	14.0
Working volume	m^3 (tuyères to stockline)	2430	1580	3930	1820	3360
Number of tuyères		28	20	42	N.A.	36
Input details (all per tonne of pig iron)						
Iron ore						
as sinter	kg	610	0	1290	1160	1500
as pellets	kg	850	1330	40	45	150
as ore	kg	0	70	270	350	0
other$^{(a)}$	kg	50	0	0	30	0

A Brief Description

Flux(b)	kg	120	180	0	0	0
Dry coke	kg	415	440	405	430	420
Blast	Nm^3 (c)	1220	1690	1040	1040	1020
Oxygen enrichment	Nm^3	20	0	10	30	15
Oil	kg	75	120	55	40	60
Moisture	kg	35	30	12	12	15
Blast temperature	K	1360	1340	1550	1330	1520
Tuyère pressure	atmospheres, gauge	2.5	2.0	3.4	N.A.	3.4
Output details						
Pig iron temperature	K	1800	1770	1760	1780	1780
Slag production	kg per tonne of pig iron	290	210	320	300	380
Slag CaO/SiO$_2$ weight ratio		1.1	1.0	1.2	1.2	N.A.
Top-gas analysis % CO (by vol.)		20	22	22	23	N.A.
% CO$_2$		20	16	22	22	21
% H$_2$		5	5	3	3	3
Top-gas temperature	K	420	470	400	400	390
Top pressure on furnace	atmospheres, gauge	1.1	0.3	2.2	N.A.	1.7

(a) Prereduced ore, scrap, millscale, etc. (b) Limestone, dolomite, BOF slag. (c) 273 K, 1 atmosphere. N.A., not available.

The Iron Blast Furnace

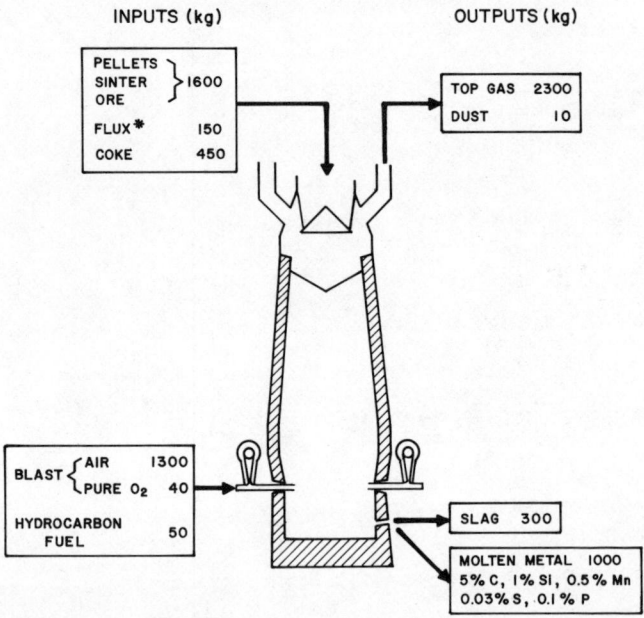

Fig. 1.3. Representative materials balance for a large blast furnace. Slag and gas compositions are given in Section 1.2.

*The principal components of the flux are CaO and MgO which are charged in prefluxed sinter or as limestone and dolomite; occasionally in steelmaking slag.

    (ii) 1–3-cm chunks of sinter produced from ore fines; and (iii) sized (1–5 cm) direct shipping ore. The charge of a particular blast furnace may contain one, two or all three of these iron oxide forms. Small charges of steelmaking slag and millscale also provide iron units.

  (b) *Metallurgical coke* (90% C, 10% ash, 0.5–1% S, dry basis: 5–10% H_2O): which supplies most of the reducing gas and heat for ore reduction and smelting. Metallurgical coke is produced by heating mixtures of powdered caking coal (25–30 wt.% volatiles content) in the absence of air. This causes the volatiles to be distilled from the coal to give a porous coke which is (i) reactive at high temperatures and (ii) strong enough to avoid being crushed near the bottom of the blast furnace. This latter property is important because pieces of coke are necessary to permit uniform gas flow through the burden as it softens and melts in the lower regions of the furnace (Nakamura, 1978). Coke is charged to the furnace as 2–8-cm pieces.

(c) *CaO and MgO*: which flux the silica and alumina impurities of ore and coke to produce a low melting point ($\cong 1600$ K), fluid slag. CaO has the beneficial secondary effect of causing part of the sulphur in the furnace charge, introduced mainly as an impurity in the coke, to be removed in the slag rather than in the product iron. The CaO and MgO are charged in pre-fluxed sinter or as 2–5-cm pieces of limestone ($CaCO_3$) and dolomite ($CaCO_3:MgCO_3$); occasionally in steelmaking slag.

Raw materials introduced through the tuyères are:

(a) *Hot-blast air*: preheated to between 1200 and 1600 K and in some cases enriched with oxygen to give blast containing up to 25 vol.% O_2. The hot blast burns incandescent ($\cong 1800$ K) coke in front of the tuyères to provide heat for (i) reduction reactions and (ii) heating and melting of the charge and products. High blast temperatures ensure that the pig iron and slag products of the furnace are well above their melting points.

(b) *Gas, liquid or solid hydrocarbons*: which provide additional reducing gas (CO and H_2) for the reduction process. Fuel oil and tar are the most common additives. Natural gas and powdered coal are also used.

1.2 Products

The main product of the blast furnace, molten pig iron, is tapped from the furnace at regular intervals (or continuously in the case of very large furnaces) through one of several holes near the bottom of the hearth. A representative analysis of molten pig iron is:

Element	Composition (wt.%)
C	4 to 5 (saturated)
Si	0.3 to 1
S	0.03
P	depends on ore, up to 1
Mn	depends on ore, 0.1 to 2.5
Melting point	1400 K

The composition of the pig iron from a particular blast furnace is chosen to meet the requirements of the steelmaking plant to which the iron is being sent. It is controlled by adjusting (i) slag composition and (ii) furnace temperature (particularly in the lower half of the furnace).

The pig iron is transported in the molten state to the steelmaking plant where the impurity elements are removed to low levels by oxygen refining. In some cases (Lu, 1975) the molten pig iron is desulphurized by treatment with calcium carbide or magnesium-coke prior to oxygen refining.

Two by-products are formed by the blast furnace:

(a) *Slag* (30–40 wt.% SiO_2, 5–15% Al_2O_3, 35–45% CaO, 5–15% MgO, 0–1% $Na_2O + K_2O$, 1–2½% S). The slag contains very little iron oxide which is indicative of the excellent reducing efficiency of the furnace. The composition of the slag is chosen (Thom, 1977): (i) to remove SiO_2 and Al_2O_3 in a fluid slag; (ii) to absorb K_2O and Na_2O (alkalis) which otherwise tend to build up in the furnace;* (iii) to absorb sulphur rather than have it dissolve in the iron product; and (iv) to control the silicon content of the metal (Shimada, 1976). A 'slag basicity' ratio:

$$\frac{\text{wt.\% } CaO + \text{wt.\% } MgO}{\text{wt.\% } SiO_2 + \text{wt.\% } Al_2O_3}$$

of between 1.1 and 1.2 appears to best meet these four slagging requirements (Ashton, 1974; Thom, 1977). Solidified blast-furnace slag is used (Emery, 1975) in the manufacture of commercial concrete and aggregate.

(b) *Gas*: which leaves through the gas-collection system at the top of the furnace. A typical modern furnace top gas composition is roughly 23 vol.% CO, 22% CO_2, 3% H_2, 3% H_2O, 49% N_2;

---

*K_2O and Na_2O enter the furnace in both coke and ore. They are partially reduced to K and Na vapour near the bottom of the furnace (Lowing, 1977) and this vapour subsequently rises to the cooler parts of the furnace where a portion reoxidizes to become entrapped in solid form in the descending burden. The process becomes cyclic and it leads to an accumulation of potassium and sodium compounds in the furnace. Physical symptoms of this buildup are a restriction of gas flow through the burden and an erratic descent of charge. Partial reduction of SiO_2 to SiO vapour is thought to cause similar problems (Elliott, 1977).

A Brief Description

equivalent to a net combustion energy of about 4000 kJ per Nm^3 (about one-tenth that of natural gas). After removal of dust, this gas is burnt in auxiliary stoves to heat the air blast for the furnace. The dust is agglomerated by sintering or briquetting and recharged to the furnace or stockpiled for future use.

1.3 Operation

Operation of the blast furnace consists of periodic charging of solids through the top; continuous or periodic tapping of liquid products from the bottom; continuous injection of hot blast with some hydrocarbons through the tuyères; and continuous removal of gas and dust.

Most of the operating procedures (e.g. charging, blowing, fuel injection) are carried out mechanically under automatic control, and modern blast furnaces are extensively equipped with continuous monitoring devices (Hashimoto, 1977). Typical continuously monitored process variables are:

temperature: hot blast, cooling water, shaft wall, top gas;
pressure: blast, furnace interior at several levels, top;
flow rates: blast (each tuyère), tuyère injectants (each tuyère), cooling water.

In addition, iron and slag temperature and composition are determined intermittently during tapping.

Burdening of the furnace (e.g. coke and flux quantities) and conditions at the tuyères (blast temperature, oxygen level, tuyère injectant quantity) are decided, often by computer, on the basis of these measurements. The only major manual manoeuvre is tapping of iron and slag.

The main chemical reactions in the furnace are (i) oxidation of carbon by air in front of the tuyères to give CO_2 plus heat:

$$C + O_2 \longrightarrow CO_2 \quad \Delta H^\circ_{298} = -394\,000 \text{ kJ (kg mole of } CO_2)^{-1}; \quad (1.1)$$

(ii) endothermic reaction of the CO_2 with carbon to produce CO, the principal reducing gas of the process:

$$CO_2 + C \longrightarrow 2\,CO \quad \Delta H^\circ_{298} = +172\,000 \text{ kJ (kg mole of } CO_2)^{-1}; \quad (1.2)$$

and (iii) reduction of iron oxides to give metallic iron:

$$Fe_{0.947}O + CO \longrightarrow 0.947Fe + CO_2$$
$$\Delta H_{298}^\circ = -17\,000 \text{ kJ (kg mole of CO)}^{-1}, \quad (1.3)$$

$$1.2Fe_3O_4 + CO \longrightarrow 3.8Fe_{0.947}O + CO_2$$
$$\Delta H_{298}^\circ = +50\,000 \text{ kJ (kg mole of CO)}^{-1}, \quad (1.4)$$

$$3Fe_2O_3 + CO \longrightarrow 2Fe_3O_4 + CO_2$$
$$\Delta H_{298}^\circ = -48\,000 \text{ kJ (kg mole of CO)}^{-1}. \quad (1.5)$$

These reactions are discussed in detail in Chapters 2 and 3.

The process never stops and a blast-furnace campaign continues 5 to 8 years before refractory wear forces a shutdown. Minor perforations in the shaft can be patched without shutting down the furnace. The furnace can be 'banked' with an all-coke charge to keep it hot during temporary production stoppages.

The internal operation of a blast furnace is extremely stable for reasons which are discussed in Chapter 3. The inception in 1828 of the use of hot blast to provide a high temperature in the hearth was the last step in eliminating any instability in the furnace. This inherent stability is one factor which has led to the pre-eminence of the blast-furnace process for producing iron.

The rate of pig-iron production in any given furnace is determined by the rate at which oxygen, as air or air plus pure O_2, is blown into the furnace. High blast rates lead to: (i) rapid combustion of coke in front of the tuyères, (ii) a rapid rate of CO production and, as a consequence, (iii) a high rate of iron reduction. 1.3 to 2.2 tonnes of blast air are required per tonne of iron.

There is a maximum rate at which air can be blown into the furnace (Lowing, 1977). Above this rate, the furnace gases tend to ascend through open 'channels' in the solid charge rather than in an evenly distributed flow pattern and this causes the reducing gases to pass through the furnace incompletely reacted. The net result is that the carbon in the charge is used inefficiently. Excessive ascent velocities also prevent newly melted iron and slag from descending evenly through the bosh ('flooding') which may lead to an uneven descent of solid charge and to an erratic furnace operation.

Typical operational blast rates are 40–50 $Nm^3\ min^{-1}$ of blast per m^2 of hearth area. A blast furnace can operate down to about 70% of its normal capacity without any deliterious effects. Below this rate of production the charge begins to react unevenly.

1.4 Improvements in productivity

The productivity of the blast furnace has continually increased from the time of its inception until today when furnaces of 12 000-tonnes-per-day capacity have been built. Much of this improvement has been due to larger furnaces of improved design (better charge distribution machinery, better cooling systems, more resistant refractories) but changes in the physical and chemical characteristics of the input materials and altered conditions inside the furnace have also led to greatly improved performances.

The most important factor in increasing blast-furnace productivity has been the switch from unsized ore to sized, evenly distributed, sinter and pellet burdens. Uniform burdens permit furnace gases to rise rapidly through the charge without causing channeling and this, in turn, allows the furnaces to be operated at high blast rates (i.e. at high production rates). In fact the productivities of older furnaces have been almost doubled by the use of sized sinter and pellets. An added benefit has been decreased losses of ore as dust in the furnace gases.

Injection of hydrocarbon liquids, gases or solids through the tuyères has also led to improved furnace productivities by lowering the requirement for coke in the solid charge. This leaves more space in the stack for ore and it leads to a greater reduction capacity. Until recently the price of coke has been high relative to the prices of fuel oil and natural gas per unit of reducing capability or enthalpy, and injection of these fuels through the tuyères has resulted in direct cost savings. However, recent increases in the prices of natural gas and fuel oil have lowered this direct-cost advantage.

Hand in hand with hydrocarbon injection has come the use of hotter blast. Hot-blast air offsets cooling due to the entry of cold hydrocarbons into the tuyère zone of the furnace and it also adds enthalpy to the system. Blast temperatures in modern installations are typically 1300 to 1600 K.

Enrichment of the air blast with pure oxygen has also been useful in improving blast-furnace performance. The oxygen replaces a portion of the air requirement and it lowers the quantity of nitrogen passing through the furnace. This partial elimination of nitrogen:

(a) increases the flame temperature in front of the tuyères thus permitting increased injection of cold hydrocarbons;
(b) permits an increased rate of CO production and consequently an increased rate of ore reduction without increasing the total rate of gas flow through the furnace.

The latter effect (b) can be used to raise the production rate of the furnace without causing channeling and flooding.

Another important improvement has been pressurization of the blast furnace up to 3 atmospheres gauge at the furnace top, and many furnaces are now operated in this way (Shimada, 1976). High-pressure operation permits an increased mass-flow rate of gas through the blast-furnace charge without an increase in gas velocity. This in turn permits a greater throughput of reducing gas without a decrease in gas/solid reaction time and it leads to an increased rate of iron production. High furnace pressure is obtained by throttling the gas exit from the furnace. The furnace structure must be strengthened for this type of operation.

1.5 Blast-furnace Costs

The previous sections have indicated that the blast furnace is a large installation with a very high production rate of molten pig iron. A single blast furnace (i.e. 5000 tonnes of iron per day) costs in the order of 100×10^6 U.S. dollars (Berczynski, 1977) including blowers, hot-blast stoves, materials-handling equipment, dust collection and effluent control. This high capital cost per unit is the biggest disadvantage of the process.

Direct operating expenses per tonne of Fe are estimated in Table 1.2, which shows that the cost of molten iron is approximately 100 U.S. dollars (1978) per tonne of contained Fe, not including investment expenses.

1.6 Summary

The iron blast furnace is an efficient device for producing large quantities of molten iron, ready for refining to steel. Its principal advantages are

A Brief Description

TABLE 1.2.

Direct Manufacturing Costs (1978) for Producing Molten Iron, 93% Fe, in a Blast Furnace. The costs reported here are at the 'study estimate' level, i.e. ±30%. Investment expenses are not included.

Iron-ore pellets (65% Fe)	1.5 tonnes @ $20 per tonne	$30
Coke	0.4 tonnes @ $100 per tonne	$40
Limestone	0.2 tonnes @ $10 per tonne	$2
Hydrocarbons (through tuyères)	0.1 tonnes @ $100 per tonne	$10
Operating cost (blast compression, labour, utilities, maintenance)		$10
Total (per tonne of pig iron)		$92

its exceptional stability of operation and its high rate of iron production. Its only disadvantage is its large unit size and its consequently high initial capital cost.

The blast furnace utilizes its fuel efficiently and this feature plus its rate of metal production have improved greatly since the inception of the process. The improvements have been especially dramatic in recent years due to (i) uniform burdening with sized ore pellets and sinter, (ii) injection of hydrocarbons to partially replace coke and (iii) furnace pressurization. Improved refractories and cooling systems, high blast temperatures and oxygen enrichment of blast have also played an important role.

Suggested Reading

Lu, W. K. (Editor) (1977 and 1978) *Blast Furnace Ironmaking 1977* and *Blast Furnace Ironmaking 1978*, McMaster University, Hamilton, Canada.

Strassburger, J. H. (Editor) (1969) *Blast Furnace – Theory and Practice*, Gordon & Breach, New York.

References

Ashton, J. D., Gladysz, C. V., Walker, G. H. and Holditch, J. (1974) 'Alkali control at DOFASCO', *Ironmaking and Steelmaking*, 1(2), 98–104.

Berczynski, F. A. (1977) 'Blast furnace design – I.', in *Blast Furnace Ironmaking 1977*, Lu, W. K., Editor, McMaster University, Hamilton, Canada, pp. 5–89.

De Marchi, G., Fontana, P. and Tanzi, G. (1978) 'Blast furnace mathematical models based on process theory, are suitable for high accuracy design, planning of operation, policy, and process control requirements', paper presented at 107th Annual AIME Meeting, Denver, Feb. 27, 1978.

Elliott, J. F. (1977) 'Blast furnace theory', in *Blast Furnace Ironmaking 1977*, Lu, W. K., Editor, McMaster University, Hamilton, Canada, Chapter 2.

Emery, J. J. (1975) 'New uses of metallurgical slags', *C.I.M. Bulletin*, 68(764), 60–68.

Hashimoto, S., Suzuki, A. and Yoshimoto, H. (1977) 'Burden and gas distribution in the blast furnace', in *Ironmaking Proceedings*, Vol. 36, Pittsburgh, 1977, AIME, New York, p. 181.

Higuchi, M., Izuka, M., Kuroda, K., Nakayama, N. and Saito, H. (1977) 'Coke quality required for operation of a large blast furnace', Metals Society Conference, Middlesbrough, England, June 14–17, 1977, summarized in *Ironmaking and Steelmaking*, 4(5), 3–4.

Lowing, J. (1977), 'The diagnostic approach to overcoming blast furnace operational problems', in *Ironmaking Proceedings*, Vol. 36, Pittsburgh, 1977, AIME, New York, pp. 224–226.

Lu, W. K., Editor (1975) *Symposium on External Desulphurization*, McMaster University, Hamilton, Canada.

Nakamura, N., Togino, Y. and Tateoka, M. (1978) 'Behaviour of coke in large blast furnaces', *Ironmaking and Steelmaking*, 5(1), 1–17.

Shimada, S. (1976) 'Biography of a big blast furnace', *Iron and Steelmaker*, 3(5), 13–22.

Sugarawa, T., Ikeda, M., Shimotsuma, T., Higuchi, M., Izuka, M. and Kuroda, K. (1976) 'Construction and operation of no. 5 blast furnace, Fukuyama Works, Nippon Kokan K.K.', *Ironmaking and Steelmaking*, 3(5), 242.

Thom, G. W. (1977) 'Fluxes and quality control', in *Blast Furnace Ironmaking 1977*, Lu, W. K., Editor, McMaster University, Hamilton, Canada, pp. 15-3 to 15-5.

Chapter 1 Problems.* *Units, Molar Quantities, Mass Balances*

Note: $1 \text{ Nm}^3 = 1 \text{ m}^3$ of gas at 273 K, 1 atmosphere. Stoichiometric data are given on page 219.

1.1 Blast furnace input and output materials have the following approximate temperatures:

(a)	blast	2200°F
(b)	ore, coke, flux	25°C
(c)	pig iron	1500°C
(d)	slag	1600°C
(e)	top gas	250°F

Express the temperatures of these materials in Kelvin (K).

*Numerical answers, page 225.

A Brief Description

1.2 A furnace is operating with a top pressure of 2×10^5 pascals which requires that the blast air be forced into the furnace at a pressure of 3.5×10^5 pascals. The temperature of the blast is 1200 K. Calculate, for these blast air conditions, the volume (m^3) of 1 tonne of dry blast.

1.3 A useful unit for expressing blast quantity is normal cubic meters (Nm^3). What is 1 tonne of air blast equivalent to in Nm^3?

1.4 A common material from which iron is produced is beneficiated and pelletized hematite ore. Calculate the quantity of hematite ore pellets (6% SiO_2) required to produce 1 tonne of pig iron (5% C, 1% Si, 1% Mn).

1.5 In modern blast furnaces, the carbon supply rate (from coke and tuyère-injected hydrocarbons) is in the order of 420 kg per tonne of pig iron (5% C, 1% Si, 1% Mn). Express this carbon supply rate in terms of (i) kg of carbon per tonne of product Fe, (ii) kg moles of C per kg mole of product Fe.

1.6 A blast-furnace operator wishes to increase his hearth temperature by enriching the blast to 25 vol.% oxygen. What weight and volume (Nm^3) of pure oxygen must he inject per 1000 Nm^3 of dry air blast?

1.7 Prepare a brief interactive computer programme which will calculate the volume (Nm^3) and weight (kg) of pure oxygen which must be added to 1000 Nm^3 of dry air blast in order to enrich the mixture to a specified vol.% O_2.

CHAPTER 2

A Look Inside the Furnace

Chapter 1 examined the blast furnace from the outside: its size, production rate, raw materials, products, operation and costs. It showed that the blast furnace is a high productivity unit which has, as its main objective, rapid production of iron from oxide ores. This chapter is devoted to examining the operation of the furnace from the inside, i.e. from the point of view of what happens to the raw materials after they enter the furnace. Much of the information in this chapter comes from studies of quenched industrial blast furnaces (Kanfer, 1974; Nakamura, 1978). Dissection of these quenched units has provided invaluable information about the internal structures of actual operating furnaces.

For coherency, the approach of the chapter is to follow the gases as they rise through the furnace, i.e. from the time the blast enters through the tuyères until the top gas emerges above the stockline. An important function of the ascending gases is the transfer of heat to the descending solid charge. This causes the gases to cool during their ascent, as is indicated by the temperature profiles in Fig. 2.1. This diagram provides a useful framework for discussing blast-furnace reactions and reaction locations. It is referred to at several points in the text.

2.1 Behaviour in Front of the Tuyères

Hot blast enters the furnace through fifteen to forty tuyères around the circumference of the top of the hearth. It enters at a velocity of 200 to 300 m s^{-1} (Hashimoto, 1977) and a pressure of 2 to 4 atmospheres, the pressure being necessary to push the reducing gases through the solid burden and to overcome the 'top pressure' of the furnace.

The high velocity of the blast clears a 'raceway' of gas and rapidly

A Look Inside the Furnace

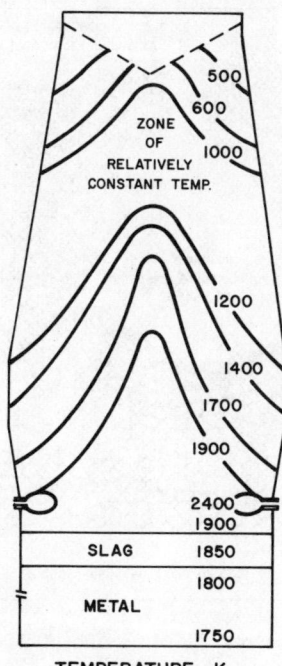

Fig. 2.1. Gas temperatures in the blast furnace as interpreted from the quenched-furnace data of Nakamura (1978). A zone of relatively constant temperature, i.e. the thermal reserve zone, is shown.

hurtling coke in front of each tuyère (Fig. 2.2). The raceways extend outwards 1 to 2 m (Nakamura, 1978) and they are easily penetrated by a bar pushed through the tuyère. They are bounded in front, at the sides and below by rather firm regions of lump coke which has bypassed oxidation during its descent through the furnace. Studies of quenched blast furnaces indicate that this coke extends downwards into the iron pool and perhaps even to the hearth bricks.

The raceways are also bounded above by lump coke, but in this case it is loosely packed due to the rapid ascent of raceway gas between the pieces. The bottom-most pieces of coke in this zone periodically fall into the raceways to be consumed by the incoming air, and hence the whole bed is always gradually moving down to be resupplied at the top with coke

18   The Iron Blast Furnace

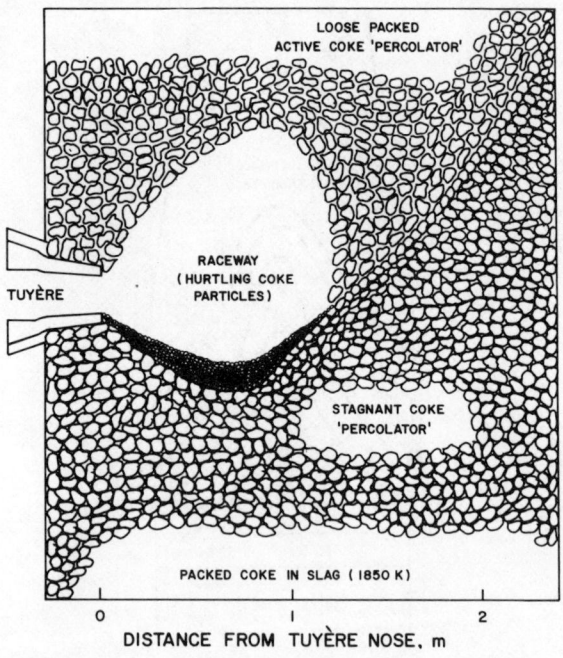

Fig. 2.2. Schematic view of tuyère region of the blast furnace as interpreted from quenched-furnace data (Nakamura, 1978). Blast-furnace tuyères are normally 15 to 20 cm I.D. at the nose and they protrude some 30 cm into the furnace.

from above. The main physico-chemical process in this region is transfer of heat from the ascending raceway gases to the descending pieces of coke and droplets of iron and slag.

2.2 Reactions in the Hearth, Tuyère Raceways and Bosh

The above discussion indicates that almost all of the solid material in the hearth and bosh is coke. Liquid iron and slag percolate through this coke to form pools in the bottom of the hearth, to be periodically or continuously tapped from the furnace. During this percolation reduction is finalized, the iron becomes saturated with carbon, and $(CaO)_3 \cdot P_2O_5$,

A Look Inside the Furnace

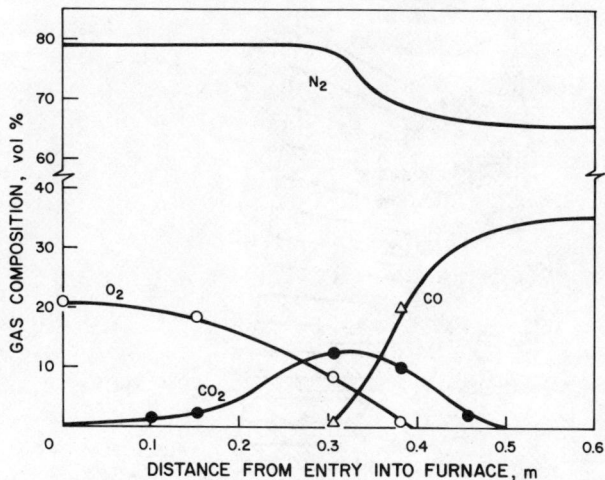

Fig. 2.3. Composition of the gas directly in front of an experimental blast-furnace tuyère (Brunger, 1970). Early combustion of coke to CO_2 and subsequent reaction of CO_2 with coke to form CO are evident.

MnO and SiO_2 are partially reduced to become impurities (Mn, P, Si) in the metal.

When oxygen first hits coke in the tuyère raceways it reacts immediately to form CO_2, as is shown in Fig. 2.3. This CO_2 then reacts further with coke to form CO by the reaction:

$$CO_2 + \underset{\text{coke}}{C} \longrightarrow 2CO \quad \Delta G°_{1800\,K} = -142\,000 \text{ kJ (kg mole of } CO_2)^{-1} \quad (1.2)$$

which, at the temperatures of the hearth and lower bosh (1800–2400 K), goes almost to completion. The tuyère gases then rise through the 'active coke' zone, Fig. 2.2, transferring heat to the descending coke and drops of iron and slag as they pass.

2.3 The Fusion Zone

Examinations of quenched blast furnaces have shown clearly that the region of loose packed coke above the raceways is bounded on top by a

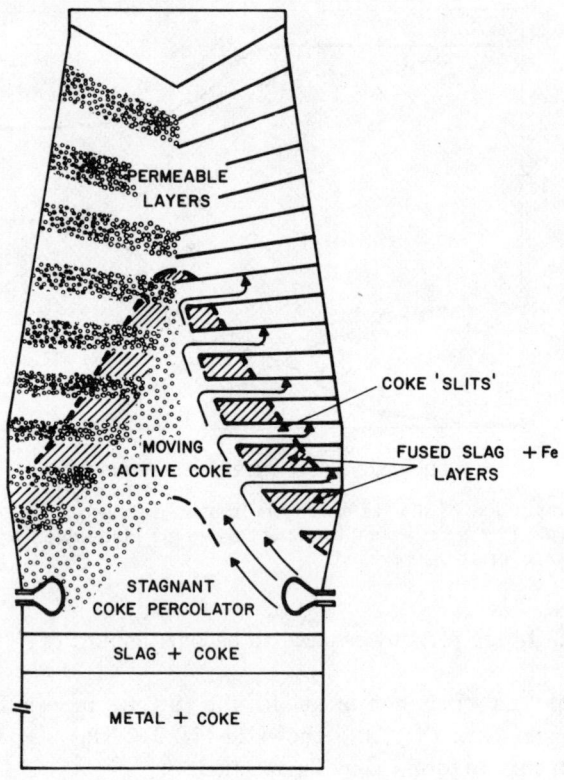

Fig. 2.4. Sketch of the internal structure of a large blast furnace (Nakamura, 1978). The most notable feature is the fusion zone with its alternate layers, ½ to 1 m thick, of (i) coke and (ii) fused slag and iron. Distribution of ascending gas through the coke 'slits' is indicated.

'fusion zone' (Fig. 2.4) consisting of alternate layers of (i) coke and (ii) softening and melting gangue, flux and iron, the layered structure having persisted from the original charging sequence.

This inverted 'U'-shaped region is important for several reasons:

(a) it, and the gas pressure below it, tend to support the furnace burden;

(b) its coke layers tend to distribute reducing gas radially across the furnace.

The latter effect arises because the softened and partially melted gangue, flux and iron are virtually impervious to gas flow so that the ascending gases must pass horizontally through the coke 'slits' in order to pass into the top of the furnace.

Since all the bosh gas must pass through the coke slits in the fusion zone, it is imperative that the coke in this region offer as little resistance as possible to the gas flow. This requires that the coke still remain as substantial lumps well down in the furnace and it is the main reason that coke must be able to avoid crushing and breaking during its descent through the furnace.

The main physical process in the fusion zone is the melting of metal and slag, making use of the heat in the ascending bosh gas. The slag at this point consists mainly of gangue and flux oxides, i.e. it does not yet contain coke ash which is for the most part released in the tuyère zones. It is almost devoid of iron oxide (Nakamura, 1978) by the time it is fully molten (i.e. at the lower edge of the fusion zone), but in any event, any iron oxide which it might contain will be fully reduced during its descent through the coke percolators.

2.4 Reduction above the Fusion Zone

The iron-bearing material in the fusion zone is principally metallic iron. Above this zone the burden begins to include solid iron oxide, specifically wustite, $Fe_{0.947}O$,* and thus at this point the burden consists of alternate layers of (i) coke and (ii) solid gangue and flux oxides, solid $Fe_{0.947}O$ and solid iron.

The gas entering this mixed burden region has risen directly from the coke bed beneath the fusion zone so that its carbonaceous component is virtually all CO. Two cyclic reactions take place in this mixed burden

---

*As used in this text, wustite is $Fe_{0.947}O$. This is the predominant wustite phase under the reducing conditions of the blast furnace.

region

reduction: $CO + Fe_{0.947}O \longrightarrow 0.947\, Fe_s + CO_2$

$$\Delta H^{\circ}_{298} = -17\,000 \text{ kJ (kg mole of CO)}^{-1} \qquad (1.3)$$

coke gasification: $CO_2 + C \longrightarrow 2\,CO$

$$\Delta H^{\circ}_{298} = +172\,000 \text{ kJ (kg mole of } CO_2)^{-1} \qquad (1.2)$$

as is schematically visualized in Fig. 2.5.

The coke gasification reaction (1.2) is highly endothermic, and it causes rapid cooling of the ascending gases. Reduction Reaction (1.3) is slightly exothermic but its heat release does not compensate for the cooling effect

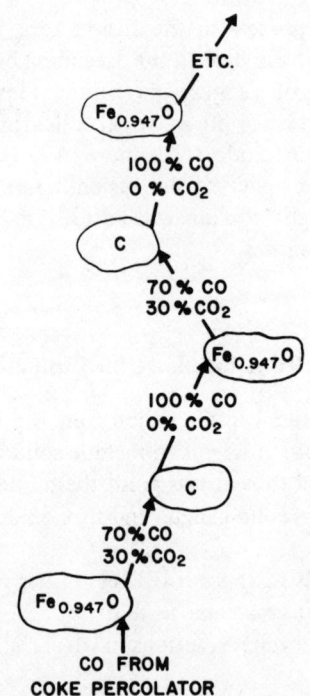

Fig. 2.5. Schematic view of cyclic reduction and coke gasification just above the fusion zone. The gas compositions are based on a reaction temperature of 1200 K and they refer to the carbonaceous components only. Solid Fe is produced by the reduction step.

of Reaction (1.2), with the net result that the temperature of the ascending gas falls markedly in this region.

2.5 Kinetics of the Coke Gasification Reaction

As will be noted in detail in Chapter 3, the rate of Reaction (1.2), coke gasification, slows markedly as temperature decreases. The result is that coke gasification comes to a virtual halt below about 1200 K which means that the cyclic reaction scheme visualized in Fig. 2.5 cannot take place below this temperature.

Thus, once the rising gases have been cooled below 1200 K, little more CO is regenerated. All subsequent reduction further up the shaft relies upon CO produced beneath the 1200 K isotherm.

2.6 Reactions in Regions above the 1200-K Isotherm

As the gas continues its ascent above the 1200-K isotherm, the CO component continues to react with wustite to form solid iron and CO_2 thereby approaching equilibrium for the reaction:

$$CO + Fe_{0.947}O \longrightarrow 0.947Fe_s + CO_2$$
$$\Delta G°_{1200 \text{ K}} = +8150 \text{ kJ (kg mole of CO)}^{-1}. \tag{1.3}$$

At 1200 K, the equilibrium pCO/pCO_2 ratio for this reaction is 2.3 and the carbonaceous portion of the gas would contain, at equilibrium, 70% CO and 30% CO_2 independent of furnace pressure and the concentrations of other gases (N_2, H_2, H_2O) in the shaft.

Reaction (1.3) is slightly exothermic and as a result the gases do not cool during their ascent through this region. This is the bottom portion of the constant temperature or 'thermal reserve' zone of the furnace.

2.7 Reduction of Higher Oxides

The rising gas eventually becomes too weak in CO to reduce significantly more wustite to iron. However, it is still strong enough to reduce

Fe_3O_4 to wustite by the reaction:

$$CO + 1.2Fe_3O_4 \longrightarrow 3.8Fe_{0.947}O + CO_2$$

$$\Delta G°_{1200K} = -8000 \text{ kJ (kg mole of CO)}^{-1} \quad (1.4)$$

for which the equilibrium pCO/pCO_2 ratio at 1200 K is 0.43 and the equilibrium proportions of CO and CO_2 are 31% and 69% respectively.

For the blast furnace to be operating at a steady state, the amount of wustite produced by Reaction (1.4) must, of course, be the same as the amount of wustite reduced by Reaction (1.3); and there is in fact more than enough CO rising from the wustite/Fe zone to accomplish this purpose. This is because:

(a) each mole of CO converted to CO_2 by reaction with Fe_3O_4 produces 3.8 moles of $Fe_{0.947}O$;
(b) the CO concentration required for $Fe_3O_4/Fe_{0.947}O$ reduction is much less than that required for $Fe_{0.947}O/Fe$ reduction, i.e. 31% versus 70%.

This excess of CO results in:

(a) creation of a vertical region in the furnace where the higher oxides have already been reduced to wustite (Kanfer, 1974) but where the gas cannot reduce significant quantities of wustite to iron;
(b) restriction of the unreduced higher oxides to a small height near the top of the furnace, about the top quarter of the shaft. This zone is only of sufficient vertical depth for its wustite production rate to equal the rate of wustite reduction lower in the furnace, i.e. it is shallow enough so that the CO passes through the zone only partially reacted.

The region where the iron-bearing material is virtually all wustite is referred to as the 'chemical reserve zone'. Because very little reaction takes place in this zone, it is also a region of roughly constant temperature. It forms, in fact, the top portion of the 'thermal reserve zone'.

Hematite reduction:

$$CO + 3Fe_2O_3 \longrightarrow 2Fe_3O_4 + CO_2$$

$$\Delta G°_{1200K} = -105\,000 \text{ kJ (kg mole of CO)}^{-1} \quad (1.5)$$

takes place along with Reaction (1.4) in the top quarter of the shaft. Reaction (1.5) can proceed at very low pCO/pCO_2 ratios ($< 10^{-4}$) so that the gas could react almost completely to CO_2 if enough Fe_2O_3 were present and if enough reaction time were available.

2.8 The Top Quarter of the Shaft and the Exit Gas

The exit gas from an efficiently operating blast furnace was shown in Chapter 1 to be in the range of 21–25% CO, 20–22% CO_2 (remainder H_2, H_2O, N_2) which has a pCO/pCO_2 ratio intermediate between the equilibrium ratios for wustite reduction to iron, Reaction (1.3), and Fe_3O_4 reaction to wustite, Reaction (1.4). This is consistent with the statement that there is more CO rising from the wustite/Fe zone of the blast furnace than can be used for the reduction of higher oxides to wustite.

The situation in the upper quarter of the shaft is, therefore, that Fe_2O_3 is being added at the top while medium-strength reducing gases are entering at the bottom. It is a zone of mixed oxides, wustite, Fe_3O_4 and Fe_2O_3, where reduction of higher oxides to wustite is taking place. In fact, individual pellets of ore or pieces of sinter may contain all three oxides.

The temperature of the gas falls rapidly as it passes through this top zone due to (i) cooling by the incoming solids; (ii) evaporation of moisture; and (iii) the net endothermic nature of reducing higher oxides to wustite, i.e.

$$CO + 1.12Fe_2O_3 \longrightarrow 2.36Fe_{0.947}O + CO_2$$
$$\Delta H^\circ_{298} = 13\,000 \text{ kJ (kg mole of CO)}^{-1}. \tag{2.1}$$

2.9 Residence Times

The above discussions have given no concept as to the reaction times available for the solids and gases to react. The space velocities* of blast furnace gases are in the order of $2–5$ m s^{-1} while their actual velocities are

---

*Volumetric flowrate at temperature ÷ cross-sectional area of empty furnace.

30–60 m s^{-1}. Gas residence times are, therefore, in the order of only 1 sec and hence it is imperative that conditions which lead to high reduction rates (e.g. good gas/solid contact) be provided for efficient blast-furnace operation.

The descent rate of the solids is much slower and the average residence time of an iron atom (ore to molten pig iron) is in the order of 5 to 8 hours (Higuchi, 1974). Thus there is considerable time provided for each ore particle to react. Since each of the wustite/Fe and Fe_2O_3/Fe_3O_4/wustite reduction zones occupies about one quarter of the furnace, an ore pellet or piece of sinter has 1 to 2 hours for each of the two main reduction steps.

2.10 Burden Arrangements

Iron blast furnaces are charged with alternate layers of coke and ore plus flux. The layers are normally 0.5–1 m thick, with larger diameter furnaces tending to have the thicker layers. As is shown in Fig. 2.4, the layered structure persists right down to the fusion zone.

Considerable attention has been paid in recent years to even distribution of the layers across the furnace, the principal objective being to produce uniform gas-flow conditions in the stack. Such uniformity of flow avoids localized regions where the furnace gas might rise so quickly through the burden that:

(a) its reducing capacity is not fully used,
(b) its heat is not efficiently transferred to the charge.

Controlled layering can also be used to deflect hot gases from the furnace shell where excessive temperatures might shorten refractory life.

Even charge distribution is obtained in bell-charged (Fig. 1.2) furnaces by rotating the bottom bell and by providing movable deflector plates (i.e. movable armour, Berczynski, 1977) at the furnace throat. An important development of the 1970s has been the installation of Paul Wurth bell-less tops, Fig. 2.6, on new and rebuilt blast furnaces. With this type of top the charge is distributed selectively over the entire cross-section of the furnace throat by means of a manoeuvrable chute. The bell-less top gives the furnace operator great selectivity in his distribution of burden and excellent control of gas flows in the upper shaft (Heynert, 1978).

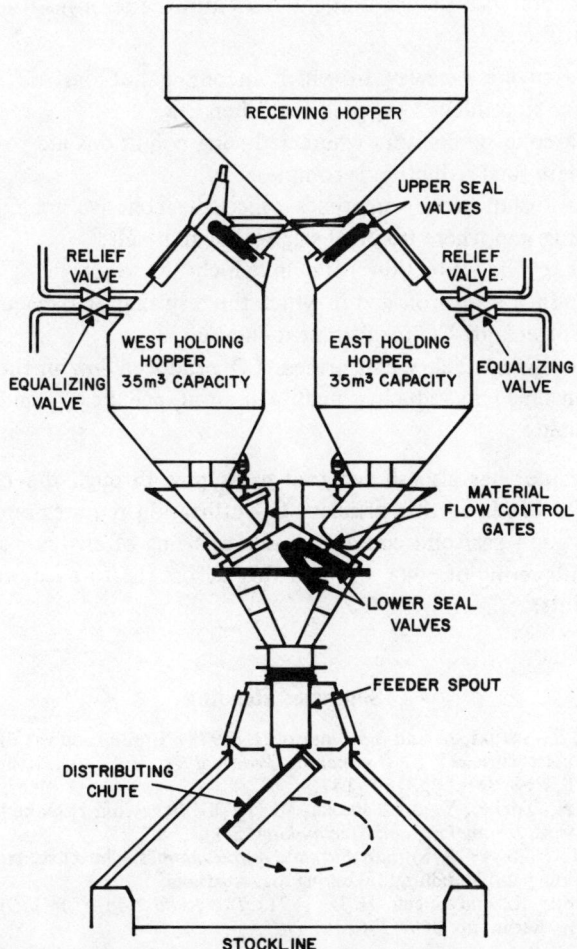

Fig. 2.6. Paul Wurth top for controlled distribution of blast-furnace burden across the furnace throat. The most significant features of the system are its manoeuvrable charging chute and its twin sets of sealing valves. The receiving hopper moves horizontally to feed one holding hopper at a time.

2.11 Summary

This chapter has shown that the blast furnace arranges itself into five important zones:

(a) the tuyère raceway in which incoming hot-blast air reacts with coke to produce CO_2 plus heat, then CO;
(b) the coke percolators where reducing conditions are strongest and where final reduction is completed;
(c) the fusion zone, sometimes called the cohesive zone, where slag forms and where iron and slag soften and melt;
(d) the cyclic reduction zone in which CO reacts with wustite to produce solid iron and in which the resulting CO_2 reacts with coke to regenerate CO for further reduction;
(e) the upper stack where excess CO produced low in the furnace is consumed in reducing wustite to iron, and Fe_3O_4 and Fe_2O_3 to wustite.

The chapter has also shown that gases pass through the furnace in a matter of seconds so that efficient CO utilization requires rapid reaction rates and good gas/solid contact. Accurate sizing of charge materials and controlled layering of coke, ore and flux across the furnace provide these requirements.

Suggested Reading

Hashimoto, S., Suzuki, A. and Yoshimoto, H. (1977) 'Burden and gas distribution in the blast furnace', in *Ironmaking Proceedings*, Vol. **36**, Pittsburgh, 1977, AIME, New York, pp. 169–187.

Nakamura, N., Togino, Y. and Tateoka, M. (1978) 'Behaviour of coke in large blast furnaces', *Ironmaking and Steelmaking*, 5(1), 1–17.

Standish, N. (Editor) (1975) *Blast Furnace Aerodynamics*, The Australasian Institute of Mining and Metallurgy, Wollongong, Australia.

Von Bogdandy, L. and Engell, H. J. (1971) *The Reduction of Iron Ores*, Springer-Verlag, Berlin, pp. 289–296.

References

Berczynski, F. A. (1977) 'Blast furnace design – I', in *Blast Furnace Ironmaking, 1977*, Lu, W. K., Editor, McMaster University, Hamilton, Canada, pp. 5–41 to 5–53.

Brunger, R. (1970) 'Experimental rig to examine combustion in the blast-furnace hearth', in *Coke in Ironmaking*, Iron and Steel Institute Publication 127, London, pp. 113–121.

Hashimoto, S., Suzuki, A. and Yoshimoto, H. (1977) 'Burden and gas distribution in the blast furnace', in *Ironmaking Proceedings*, Vol. 36, Pittsburgh, 1977, AIME, New York, p. 173.

Heynert, G., Peters, K. H. and Ringkloff, G. (1978) 'Five years of experience with the bell-less top', *Iron and Steelmaker*, 5(3), 15–24.

Higuchi, M., Kuroda, K. and Nakatani, G. (1974) 'Operation of blast furnaces at Fukuyama Works, Nippon Kokan K.K.', *Iron and Steel Engineer*, 51(9), 47.

Kanfer, V. D. and Murav'ev, V. N. (1974) 'Investigation of production processes in a quenched industrial blast furnace', *Steel in the U.S.S.R.*, 4(9), 864–868.

Nakamura, N., Togino, Y. and Tateoka, M. (1978) 'Behaviour of coke in large blast furnaces', *Ironmaking and Steelmaking*, 5(1), 5–7.

Chapter 2 Problems. *Energy, Stoichiometry, Gas–Solid Equilibrium*

2.1 Industrial thermal data are often reported in British Thermal Units. Compare this unit to the kilojoule, the unit which is used throughout this text.

2.2 The coke gasification (Boudouard) reaction is represented by:

$$CO_2 + C \rightleftharpoons 2CO$$

for which $\Delta G°_{1900K}$ is $-159\,000$ kJ (kg mole of CO_2)$^{-1}$.
Calculate using this information:

(a) the equilibrium ratio: $(pCO)^2/pCO_2$ for the Boudouard reaction at 1900 K;

(b) the composition (vol.% CO, CO_2, N_2) of gas rising through the 'active coke zone' (1900 K, 2×10^5 pascals total pressure) of a blast furnace if $C/CO/CO_2$ equilibrium is attained;

(c) the effect on your answer to part (b) of increasing the furnace pressure by 10^5 pascals.

Assume in parts (b) and (c) that (i) all oxygen in the gas originates from dry air blast and that (ii) pO_2 in the coke zone is negligible.

2.3 Prepare a brief interactive computer programme which will calculate the composition of the 'active coke zone' gas specified in Problem 2.2 for given values of:

(a) temperature;
(b) total pressure;
(c) volume percent O_2 in blast air.

Use the programme to confirm your answers to Problems 2.2(b) and 2.2(c).
The standard free energy for the coke gasification reaction may be represented by:

$$\Delta G° = 163\,000 - 169T \quad \text{kJ (kg mole of } CO_2)^{-1}$$

The Iron Blast Furnace

2.4 Space velocity is a useful variable for comparing gas flows in furnaces of different sizes. Calculate, therefore, the space velocity of gas (1900 K, 2×10^5 pascals) rising through the 'active coke zone' of Fukuyama number 5 furnace (data, Table 1.1).

Assume that:

(a) the diameter of the furnace in the region of the active coke zone is 15 m;
(b) the blast volume given in Table 1.1 is for moist blast;
(c) Fukuyama oil is 88 wt.% C, 12 wt.% H;
(d) oxide reduction is negligible in the coke zones.

Hint: all oxygen in the blast (from air, O_2 and H_2O) rises through the active coke zone as CO.

CHAPTER 3

Thermodynamics of the Blast-Furnace Process: Enthalpies and Equilibria

Chapter 2 showed that the iron blast furnace is: (i) a counter-current gas/solid heat exchanger from tuyère raceways to stockline; (ii) a counter-current oxygen exchanger from fusion zone to stockline. These exchange characteristics are shown clearly by the behaviour of the furnace gas which, as it ascends,

(a) transfers heat to the charge for heating and melting and for endothermic reactions, particularly gasification of coke, Reaction (1.2);
(b) removes oxygen from the iron oxide charge by forming CO_2 from CO.

The counter-current nature of these exchanges leads to the high chemical and thermal efficiencies of blast furnaces and to high rates of iron production.

This chapter considers the thermal requirements of the blast-furnace process and it examines the existence of distinct active and reserve (inactive) regions in the furnace.

3.1 Enthalpy Requirements in the Blast Furnace

The blast furnace has two basic enthalpy requirements:

(a) the enthalpy supply must meet the demands of the overall process;
(b) there must be enough high temperature enthalpy to:
  (i) melt and superheat the iron and slag;
  (ii) maintain a large reduction zone (temperature > 1200 K) in

which Fe is reduced rapidly from wustite and in which CO is regenerated rapidly to perform more reduction.

This last condition is necessary for the furnace to produce iron at a high rate.

Requirements (a) and (b) are independent variables, i.e. each could be satisfied without satisfying the other, and hence both must be met. For example, a large quantity of room-temperature blast, burning a large quantity of coke, would satisfy the overall enthalpy requirements of the furnace but the high hearth and reduction zone temperature requirements would not be met.

From the enthalpy balance point of view (but not a mechanistic point of view), blast-furnace reduction reactions may be written:

$$3/2C + 1/2Fe_2O_3 \longrightarrow Fe_s + 3/2CO$$
$$\Delta H^{\circ}_{298} = +247\,000 \text{ kJ (kg mole of Fe)}^{-1}, \tag{3.1}$$

$$3/4C + 1/2Fe_2O_3 \longrightarrow Fe_s + 3/4CO_2$$
$$\Delta H^{\circ}_{298} = +118\,000 \text{ kJ (kg mole of Fe)}^{-1}, \tag{3.2}$$

$$3/2H_2 + 1/2Fe_2O_3 \longrightarrow Fe_s + 3/2H_2O$$
$$\Delta H^{\circ}_{298} = +50\,000 \text{ kJ (kg mole of Fe)}^{-1}. \tag{3.3}$$

which shows that irrespective of the nature of the gaseous product, there is a net enthalpy deficit for the overall reduction process.

This enthalpy deficit (plus heating, melting and furnace loss requirements) is made up by:

(a) introducing hot blast air through the tuyères;
(b) oxidizing carbon with this hot air to form CO (the raceway product) plus heat, i.e.

$$C + \tfrac{1}{2}O_2 \underset{\text{(blast)}}{\longrightarrow} CO \quad \Delta H^{\circ}_{298} = -111\,000 \text{ kJ (kg mole of C)}^{-1}. \tag{3.4}$$

Quantitative incorporation of these data into the mathematical description of the blast furnace process is presented in Chapter 5 onwards.

3.2 Critical Hearth Temperature

The most critical temperature in the blast furnace is that in the hearth where the slag must be held in a molten and fluid state. The slag does not contain any iron oxide because of the highly reducing conditions in the coke percolators and hence its melting point is high ~1600 K. Including provision for superheat, the hearth must, therefore, be at about 1800 K.

This hearth temperature is obtained by providing conditions which produce a high flame temperature in front of the tuyères, namely a hot blast and the descent of hot incandescent coke to tuyère level. The principal factor establishing the tuyère flame temperature of any furnace is the temperature of the ingoing blast, the effects of which are shown in Fig. 3.1. The significant features of Fig. 3.1 are:

(a) flame temperature exceeds the critical hearth temperature (1800 K) when dry blast is preheated to 500 K or more;
(b) flame temperature increases proportionally to blast temperature.

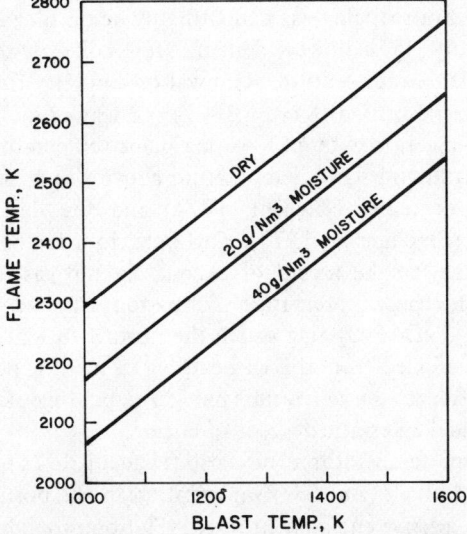

Fig. 3.1. Adiabatic flame temperature when incandescent coke, 1800 K, is combusted with air to form CO. The effects of blast temperature and moisture (normally 10–30 g of H_2O per Nm^3 of air) are shown. The definition of tuyère flame temperature used throughout this text is presented in Appendix I.

Flame temperature can also be increased by enriching the blast with oxygen which has the effect of decreasing the amount of inert nitrogen that must be heated in the tuyère zone. Two factors which lower flame temperature (and thus hearth temperature) are:

(a) humidity in the ingoing air blast (Fig. 3.1) due to the endothermic reaction:

$$H_2O + C \longrightarrow CO + H_2 \quad \Delta H^{\circ}_{298} = +131\,000 \text{ kJ (kg mole of } H_2O)^{-1}; \quad (3.5)$$

(b) hydrocarbon injection with the blast. This effect is due to the low temperature of the hydrocarbon injectant (compared with that of the incandescent coke) and to the endothermic nature of hydrocarbon dissociation.

Common blast-furnace practice is to balance off a high blast temperature and oxygen enrichment against hydrocarbon injection with the blast. This results in extra reducing gas and enthalpy being added to the furnace (and consequently in a lower demand for coke) without adversely decreasing the flame temperature. Removal of humidity from the air blast by refrigeration is also used (Kurita, 1977) as a way of increasing the level of hydrocarbon injection without lowering flame temperature.

There is a maximum flame temperature above which the permeability of the burden decreases (Higuchi, 1974) and the descent of charge becomes erratic (Stephenson, 1977). This behaviour is due to overheating in the furnace, i.e. to the ascent of excessively hot gas from the tuyère raceways, which causes premature formation of low melting point (~1400 K) CaO–FeO–SiO$_2$ slag above the normal fusion zone. The FeO is subsequently reduced from this slag causing its melting point to increase and the slag to freeze, the net results being a cementing of particles above the fusion zone and an erratic descent of charge.

Excessive flame temperatures may also (Higuchi, 1974) cause abnormally high rates of alkali (and SiO) vaporization in the bosh which in turn can lead to (i) excessive entrapment of reoxidation products in the burden and (ii) decreased burden permeability.

These overheating problems are avoided by generating an appropriate flame temperature, about 2400 K as defined in Appendix I, by manipulating blast temperature, oxygen enrichment, humidity and hydrocarbon injection as described above. Precise control of flame temperature has the added advantage that it leads to a continuous supply of constant composition metal (Dartnell, 1975).

3.3 Temperature Profiles in the Furnace: The Thermal Reserve Zone

The vertical temperature profile of gases in the iron blast furnace as determined by probe analysis indicates the presence of three distinct zones (Fig. 3.2).

(a) Bottom zone (bosh, belly, and lower shaft): in which the gases cool rapidly as they ascend.

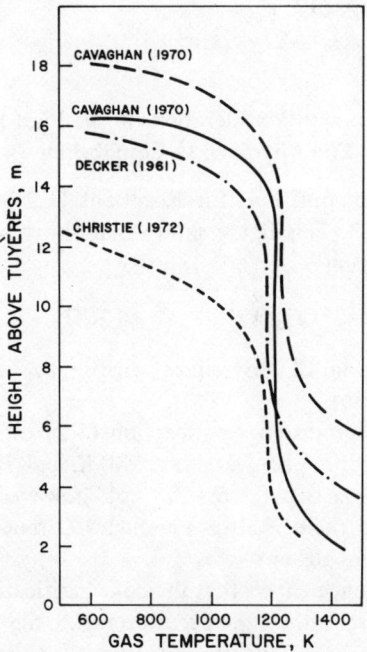

Fig. 3.2. Vertical temperature profiles in operating blast furnaces as determined by thermocouple probes (see also Lowing, 1977). The lines represent averaged data ±100 K.

(b) Middle zone: in which the temperature of the rising gases remains relatively constant (the thermal reserve zone).
(c) Top zone (top quarter of the shaft): in which the gases cool quickly during their final ascent to the stockline. The gases leaving the middle zone contain more sensible heat than is necessary to bring the incoming solids up to the thermal reserve temperature. As a consequence, (i) the top zone is thin and (ii) the gases leave the furnace 100 to 200 K above the temperature of the incoming charge.

These zones are evident in the data of Decker (1961), Cavaghan (1970) and Christie (1972) who pushed several types of temperature probes into the furnace. The thermal reserve zone is not always as clearly defined as shown in Fig. 3.2, but a zone of relatively constant gas temperature is usually found (Lowing, 1977).

3.4 Free Energy Considerations in the Blast Furnace: The Approach to Equilibrium

Equilibrium gas compositions for Reactions (1.2), (1.3) and (1.4) are represented in Fig. 3.3. This figure shows that above about 960 K the gas produced by the reaction

$$CO_2 + C \longrightarrow 2CO \qquad (1.2)$$

is sufficiently strong in CO to reduce wustite to iron but that below this temperature it is not.

Similarly the gas produced by Reaction (1.2) can reduce Fe_3O_4 to wustite at all temperatures above about 940 K and Fe_2O_3 to Fe_3O_4 at any temperature. These two factors are not, however, important in the blast furnace because there is always enough CO remaining after wustite reduction to reduce the higher oxides.

Experimental evidence shows that the coke gasification reaction closely approaches equilibrium in the coke percolators, i.e. the carbonaceous gas is virtually all CO, Fig. 2.3. However, Reaction (1.2) has a high activation energy, 360 000 kJ (kg mole)$^{-1}$ (von Bogdandy, 1971), and it slows greatly as the system cools, Fig. 3.4. The result is that at temperatures

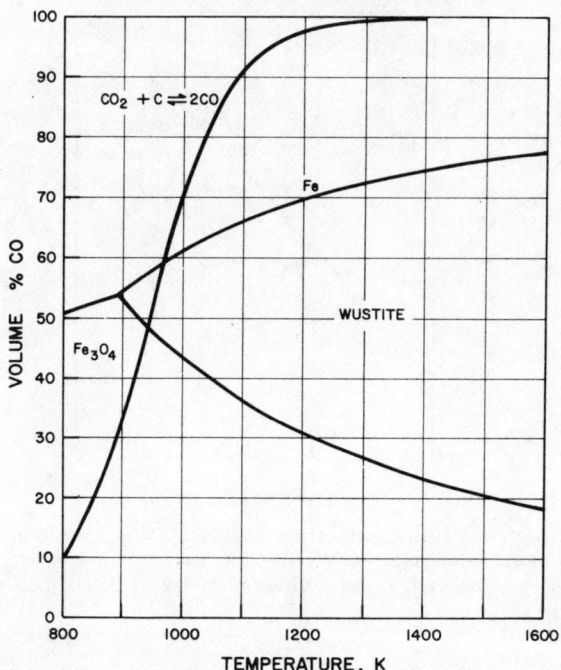

Fig. 3.3. Equilibrium gas compositions for Reactions (1.2), (1.3) and (1.4) expressed as vol.% CO in the carbonaceous portion of the gas. The Reaction (1.2) curve is based on a total carbonaceous gas pressure ($p\text{CO} + p\text{CO}_2$) of one atmosphere. Thermodynamic data: Stull, 1970.

below about 1200 K the CO_2 produced by the wustite reduction reaction

$$CO + Fe_{0.947}O \longrightarrow 0.947\,Fe + CO_2 \qquad (1.3)$$

is no longer reconverted to CO by Reaction (1.2). For this reason Reaction (1.2) is far out of equilibrium in the remaining height of the furnace.*

---

*For this same reason soot formation, the reverse of Reaction (1.2),

$$2CO \longrightarrow C + CO_2$$

does not occur significantly in the shaft even though it is thermodynamically favoured at temperatures below about 950 K.

38   The Iron Blast Furnace

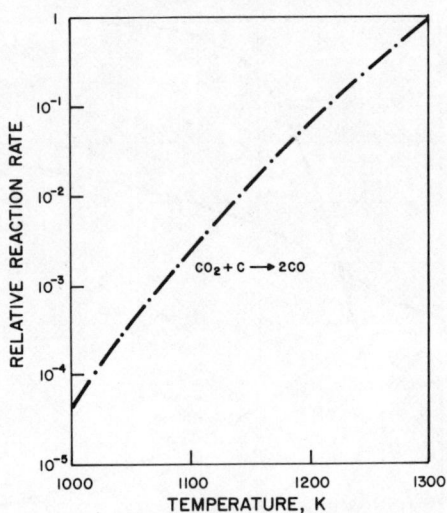

Fig. 3.4. Relative rates of coke gasification reaction (1.2) as a function of temperature. The rate has been taken as unity at 1300 K. Rates at other temperatures have been calculated on the basis of an activation energy of 360 000 kJ (kg mole)$^{-1}$ (Von Bogdandy 1971).

3.5 Gas Composition Profiles in the Furnace: The Chemical Reserve Zone

At elevations above the 1200-K isotherm, virtually no CO is produced by coke gasification. All reduction higher in the furnace than this isotherm is carried out by CO generated below the isotherm. As it rises, this CO sequentially reduces $Fe_{0.947}O$ to Fe; Fe_3O_4 to $Fe_{0.947}O$; and Fe_2O_3 to Fe_3O_4.

As was pointed out in Section 2.7, the stoichiometry of this sequence of reactions favours production of more wustite from higher oxides than is reduced to metallic iron by an equivalent amount of gas. The net result is the creation of a height in the stack, the chemical reserve zone, where the iron-bearing material is virtually all wustite (Kanfer, 1974) and where the gas composition has approached that for $Fe_{0.947}O$/Fe equilibrium. Evidence of such chemical reserve zones is provided by the data of Fig. 3.5 which clearly show the presence of a vertical region in the blast furnace

Process Enthalpies and Equilibria

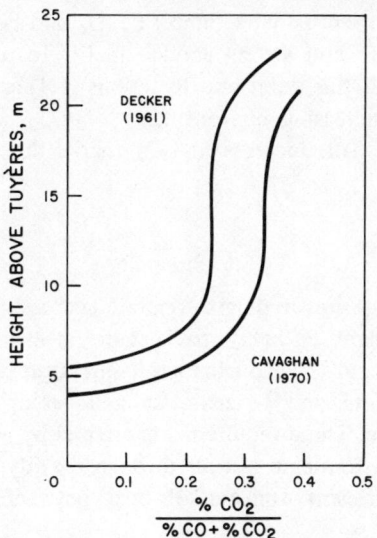

Fig. 3.5. Vertical gas composition profiles near the sidewalls of operating blast furnaces as determined by gas probes. For probe details see Lowing (1977).

where:

(a) the gas does not change in composition as it rises;
(b) the composition is near the equilibrium composition (70% CO, 30% CO_2) for Reaction (1.3) at the thermal reserve temperature, 1200 ± 100 K.

The vertical extent of the chemical reserve zone varies from furnace to furnace and such a zone may not, in fact, be present when a furnace is being 'pushed' for a high rate of production. It is likely that wustite/Fe equilibrium is not attained under such 'pushing' conditions, in which case (i) CO continues to reduce wustite until Fe_3O_4 is eventually encountered high in the shaft and (ii) a wustite-only zone with constant gas composition does not exist. Even under these conditions, however, the CO always becomes depleted near the top of the wustite/Fe region and a zone in which gas and solid compositions change only slowly is always present.

The rising gases eventually encounter Fe_3O_4 and Fe_2O_3 near the top of the shaft. They are still strong enough in CO to reduce both of these oxides to wustite at this point and Reactions (1.4) and (1.5) take place as indicated by an increasing concentration of CO_2 near the top of the furnace (Fig. 3.5). This increase in CO_2 marks the top of the chemical reserve zone.

3.6 Summary

This chapter has shown that to operate successfully the blast furnace requires (i) sufficient enthalpy to sustain its entire reduction/heating melting sequence and (ii) enough high-temperature enthalpy to keep its metal and slag fluid and to maintain a large high-temperature, rapid reduction-rate zone. These requirements are met by adding enough fuel to the furnace; by introducing hot air through the tuyères; and by creating conditions for efficient transfer of heat between ascending gas and descending burden.

The chapter has also shown that the blast furnace arranges itself into chemically and thermally active and inactive (reserve) zones as are outlined in Table 3.1. The three most significant zones are the coke percolators, the thermal reserve zone and the chemical reserve zone. These three reserves act as buffers in the furnace which smooth out any temporary perturbations (e.g. an inadequate coke supply, uneven layering) which the furnace might encounter, thereby ensuring the stable behaviour of the process.

Suggested Reading

Kitaev, B. I., Yaroshenko, Y. G. and Suchkov, V. D. (1967) *Heat Exchange in Shaft Furnaces*, Pergamon Press, Oxford.

Staib, C., Rist, A. and Michard, J. (1967) 'Control and automation of the blast furnace', in *Ironmaking Tomorrow*, Proceedings of the Autumn General Meeting of the Iron and Steel Institute, Nov. 22 and 23, 1966 (Iron and Steel Institute Publication 102), pp. 85–95.

References

Cavaghan, N. J. and Wilson, A. R. (1970) 'Use of probes in blast furnaces', *J. Iron and Steel Institute,* **208**(3), 231–246.

TABLE 3.1.

Summary of Reactions and Thermal Behaviour in Various Regions of the Iron Blast Furnace (steady-state conditions)

Zone	Chemical behaviour	Thermal behaviour
Hearth (below tuyères)	Saturation of iron with carbon and final reduction of $(CaO)_3 \cdot P_2O_5$, MnO and SiO_2. Impurities reach their final concentrations	Falling drops of metal and slag bring heat down into the hearth
Tuyère raceways	Coke and hydrocarbons are oxidized to CO_2 then CO	Large evolution of heat from combustion of incandescent coke with hot air
Active coke zone (bosh and belly)	Impurity oxides are reduced and iron absorbs carbon during percolation of metal and slag droplets. Coke bed acts as a reserve to guarantee complete reduction of iron oxides	Transfer of heat from ascending gas to descending coke and droplets of metal and slag. Gas supplies enthalpy for superheating of liquid metal and slag and for reduction of impurity metals
Fusion zone	Formation and melting of slag, final reduction of $Fe_{0.947}O$, melting of Fe	Transfer of heat from ascending gas to melting solids
Cyclic reduction zone	Wustite reduction (Reaction (1.3)) and coke gasification (Reaction (1.2))	Temperature of ascending gas falls rapidly due to absorption of enthalpy by Reaction (1.2)
Middle of shaft	Very little coke gasification occurs. Reduction of $Fe_{0.947}O$ by CO over much of the zone. Little or no reaction in higher portions due to depletion of CO (chemical reserve zone). Wustite is the iron oxide throughout this region	Steady temperature (≈ 1200 K) throughout, (thermal reserve zone). Wustite reduction to Fe requires no enthalpy
Upper quarter of shaft	Reduction of Fe_2O_3 and Fe_3O_4 to wustite	Temperature of gases decreases rapidly due to transfer of heat to cold incoming solids. Reduction reactions and evaporation of moisture also require enthalpy

Christie, D. H., Kearton, C. J. and Thomas, R. (1972) 'Practical application of mathematical models in ironmaking', in *Blast Furnace Technology, Science and Practice*, Edited by Szekely, J., Marcel Dekker, Inc., New York, pp. 115–170.

Dartnell, J. (1975) 'Consistent iron quality', *Ironmaking and Steelmaking*, 2(2), 99–100.

Decker, A. (1961) discussion of 'Use of blast furnace probes', *Proceedings of Blast Furnace, Coke Oven, and Raw Materials Conference (AIME)*, 1961, 20, 46.

Higuchi, M., Kuroda, K. and Nakatani, G. (1974) 'Operation of blast furnaces at Fukuyama Works, Nippon Kokan K.K.', *Iron and Steel Engineer*, 51(9), 47.

Kanfer, V. D. and Murav'ev, V. N. (1974) 'Investigation of production processes in a quenched industrial blast furnace', *Steel in the U.S.S.R.*, 4(9), 864–868.

Kurita, M., Yabe, S. and Shimizu, H. (1977) 'Improvements and operation of Kashima Works' No. 1 blast furnace', in *Ironmaking Proceedings*, Vol. 36, Pittsburgh, 1977, AIME, New York, p. 114.

Lowing, J. (1977) 'The diagnostic approach to overcoming blast furnace operational problems', in *Ironmaking Proceedings*, Vol. 36, Pittsburgh, 1977, AIME, New York, pp. 212–233.

Stephenson, R. L. (1977) 'Optimum blast-furnace hot-blast temperature', *Iron and Steelmaker*, 4(7), 31–33.

Stull, D. R., Prophet, H. *et al.* (1970) *JANAF Thermochemical Tables, 2nd Edition*, United States Department of Commerce, Document NSRDS-NBS 37, Washington, June 1971.

Von Bogdandy, L. and Engell, H. J. (1971) *The Reduction of Iron Ores*, Springer-Verlag, Berlin, pp. 289–296.

Chapter 3 Problems. *Flame Temperatures, Enthalpy, Kinetics*

Example flame temperature calculations: Appendix I. Data: Appendices V, VI.

3.1 By the time that the coke in a blast furnace charge has descended to tuyère level it has been heated to approximately 1800 K. This coke is combusted by the oxygen of the blast to give CO at the edge of the tuyère raceways (Fig. 2.3). The flame temperature at the edge of the tuyère raceways determines to a considerable extent the final temperature of the slag and metal and hence it is important to understand how this flame temperature is affected by furnace operating parameters.

Calculate, therefore, the adiabatic flame temperature at the edge of the tuyère raceways (overall combustion to CO and H_2) under the following conditions:

(a) dry air blast, 1200 K;
(b) dry blast enriched to 25% O_2 by adding pure oxygen, all at 1200 K;
(c) air blast containing 8 g of H_2O per Nm^3 of moist air, 1200 K;
(d) dry air blast, 1200 K, to which fuel oil (298 K, 86% C, 14% H, gross

heat of combustion = $-46\,600$ kJ per kg of oil) is added, 20 kg of oil per 1000 Nm^3 of blast air;

(e) moist air (8 g of H_2O per Nm^3 of moist blast), 1500 K, to which natural gas (CH_4, 298 K) is added, 0.1 Nm^3 of CH_4 per Nm^3 of moist air.

Ignore impurities in the coke.

3.2 A blast-furnace operator wishes to inject hydrocarbons into a new blast furnace and to make this possible he has constructed sufficient stove capacity to consistently give a 1500-K blast temperature. For an adequate hearth temperature, the flame temperature at the edge of the tuyère raceways must be maintained at 2400 K.

Calculate for these circumstances:

(a) the maximum quantity of fuel oil, 298 K (see Problem 3.1(d) for details), which could be injected into the furnace with dry blast;

(b) the maximum quantity of CH_4, 298 K, which could be injected into the furnace with moist blast (8 g of H_2O per Nm^3 of moist blast). In this case express your answer in terms of (i) kg moles of CH_4 per kg mole of O from dry blast air and (ii) Nm^3 of CH_4 per 1000 Nm^3 of moist blast.

Ignore impurities in the blast-furnace coke.

3.3 Prepare a brief interactive programme which will calculate:

(a) the tuyère-edge flame temperature generated by the injection of any given mix of moist air, oxygen and hydrocarbon fuel through the blast-furnace tuyères;

(b) the quantity of hydrocarbon fuel which must be added through the tuyères to maintain a specified flame temperature, given the temperature and moisture content of the blast and the oxygen-enrichment level.

The heat of combustion of the hydrocarbon fuel is known in both cases. Ignore impurities in the coke.

Test your programme on Problems 3.1 and 3.2.

3.4 The ratio of the rate constants for coke gasification and ore reduction (expressed in terms of moles of CO_2 produced or consumed per m^2 of surface area) is in the order of 1 at 1450 K.* What will this ratio be at (i) 1200 K; (ii) 1000 K?

Data: Activation energies:

Coke gasification	360 000 kJ (kg mole)$^{-1}$
Ore reduction	50 000 kJ (kg mole)$^{-1}$

*Von Bogdandy, L. and Engell, H. J. (1971) *The Reduction of Iron Ores*, Springer-Verlag, Berlin, p. 295.

CHAPTER 4

Blast-Furnace Stoichiometry

As the previous chapters have shown, the blast-furnace process entails:

(a) combusting coke with hot blast in front of the tuyères to produce CO plus heat;
(b) transferring oxygen from the descending iron oxide burden to the ascending CO gas;

the net result being production of molten iron. This chapter begins the development of a mathematical description of these processes.

The ultimate objective of the mathematical description is to provide equations which can be used:

(a) to calculate explicitly the quantities of coke and blast air which are required to produce iron from any given ore;
(b) to determine the effects of altering furnace operating parameters (e.g. blast temperature, hydrocarbon injection, oxygen enrichment) on these coke and blast requirements.

The development of the mathematical description is begun (this chapter) by first preparing a materials-balance equation for the process. A simplified enthalpy-balance equation is then developed (Chapter 5) and the two are combined (Chapter 6) to provide a framework for the model.

The overall furnace approach of Chapters 4, 5 and 6 is instructive as to how the blast furnace behaves, but it cannot provide an *a priori* mathematical description of how the blast furnace must be operated for any given ore and flux charge. This final goal of an *a priori* predictive model is attained by conceptually dividing the furnace, top from bottom, through its chemical reserve zone and by preparing stoichiometric and enthalpy balances for the bottom segment (Chapters 7, 8 and 9).

4.1 The Stoichiometric Development

The equations of this chapter describe steady-state operating conditions, i.e. conditions under which there is no accumulation of mass in the furnace. This permits balancing of the carbon, iron and oxygen inputs and outputs of the furnace and it provides the foundation for developing a stoichiometric equation for the process. For simplicity, the stoichiometric equations are developed on a molar basis. As will be seen from illustrative calculations in the text and from problems at the end of the chapter, it is a relatively simple matter to convert to kilograms and tonnes.

4.1.1 Elemental Balances

Under steady-state conditions the blast furnace must obey three basic elemental balance equations:

$$n^i_{Fe} = n^o_{Fe}, \qquad (4.1)$$

$$n^i_C = n^o_C, \qquad (4.2)$$

$$n^i_O = n^o_O, \qquad (4.3)$$

where n^i is the number of moles of each component entering the furnace and n^o is the number of moles leaving, *all expressed in terms of a production of one mole of useful Fe in pig iron, i.e. per mole of product Fe.*

Fe, C and O enter and leave the furnace in many forms, but initially the only forms which need be considered are:

Element	*Forms into furnace*	*Forms out of furnace*
Fe	Iron oxide	Metal
C	Coke	CO, CO_2, carbon-in-iron
O	Iron oxides, blast air	CO, CO_2

The Iron Blast Furnace

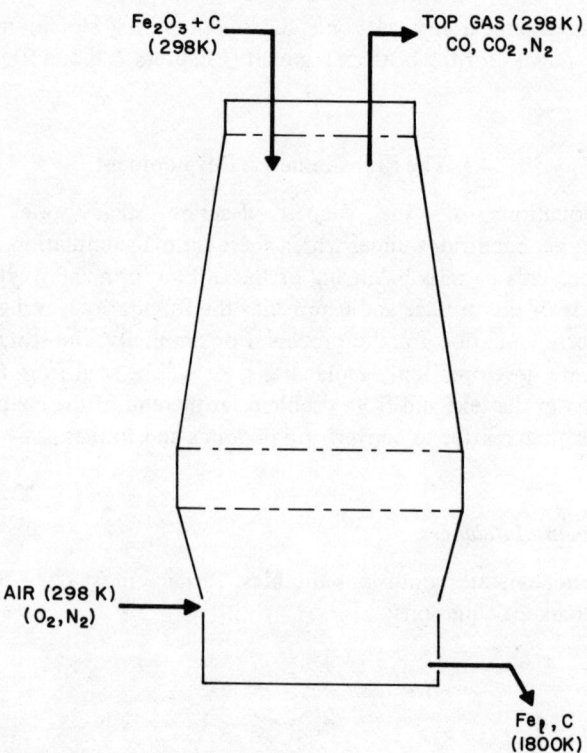

Fig. 4.1. Inputs and outputs of the iron blast furnace as simplified for the materials and enthalpy balances of Chapters 4, 5 and 6.

as is shown in Fig. 4.1. This table is based on the industrial observations that:

(a) Fe leaves the furnace almost exclusively in the blast-furnace metal,* i.e. less than 0.5% leaves in the slag (Tsuchiya, 1976);
(b) the oxygen content of blast-furnace metal is negligibly small, so that no oxygen leaves the furnace in this product;

---

*Loss of input materials as dust in top gas, 10 to 20 kg per tonne of Fe produced by the furnace, is ignored in the mass balance equations, i.e. it is assumed that all dusts are eventually recycled to the furnace.

(c) oxides which enter the furnace in gangue, flux and coke ash leave the furnace mainly in slag, so that these oxides can, for the moment, be ignored (impurity reduction and carbonate calcination are considered later).

All further equations in this chapter are based on these observations as represented by Fig. 4.1.

4.1.2 Composition Ratios

Before developing the stoichiometric equation, it is useful to introduce several variables which concisely represent C, Fe and O concentrations and quantities in the furnace inputs and outputs.

A useful variable for describing top-gas composition is, for example;

$$(O/C)^g$$

which is the molar oxygen/carbon ratio in the top gas. This variable represents concisely the mole fractions of CO and CO_2 in the carbonaceous portion of the top gas ($X_{CO}^g, X_{CO_2}^g$) to which it is related by:*

$$X_{CO_2}^g = (O/C)^g - 1,$$

$$X_{CO}^g = 2 - (O/C)^g.$$

Likewise, the composition of the iron oxide portion of the charge may be expressed as:

$(O/Fe)^x$ (moles of O per mole of Fe in the iron oxides)

which, for hematite is 3/2; magnetite, 4/3; and a mixed charge, somewhere inbetween.

Finally, the carbon content of the pig iron product may be expressed in terms of the molar ratio:

$(C/Fe)^m$ moles of C in pig iron, per mole of Fe in pig iron.

---

*Assuming (as observed) that blast-furnace gases contain no O_2. The complications introduced by H_2 and H_2O in the furnace are discussed in Section 11.2.

These ratios can also be used to concisely represent certain quantities of oxygen and carbon entering and leaving the furnace. In this context their meanings are:

$(O/C)^g$ — moles of oxygen leaving in top gas per mole of carbon leaving in top gas,

$(O/Fe)^x$ — moles of oxygen entering in iron oxides per mole of Fe entering in iron oxides,

$(C/Fe)^m$ — moles of C leaving in molten metal per mole of product Fe.

4.1.3 Material Balances

Steady-state operation of the Fig. 4.1 blast furnace is described by the following four materials balance equations, all expressed per mole of product Fe (i.e. per mole of Fe in blast-furnace metal).

(a) Iron balance

Since the only outlet for Fe from the Fig. 4.1 furnace is in the product metal, equation (4.1) becomes, per mole of product Fe,

$$n^i_{Fe} = n^o_{Fe} = 1. \tag{4.4}$$

(b) Carbon balance

Carbon leaves the furnace in (i) the top gas and (ii) the molten pig iron, i.e.

$$n^o_C = n^g_C + (C/Fe)^m \tag{4.5}$$

where n^g_C is moles of carbon leaving in the top gas and $(C/Fe)^m$ is moles of carbon leaving in the metal, both per mole of product Fe.

(c) Oxygen balances

Oxygen enters the Fig. 4.1 furnace in blast air and iron oxides, i.e.

$$n^i_O = n^B_O + (O/Fe)^x \tag{4.6}$$

where n_O^B is the moles of O brought into the furnace in blast air per kg mole of product Fe. $(O/Fe)^x$ is the oxygen input of the iron oxides per mole of their input Fe, but by equation (4.4) it can also be expressed per mole of product Fe.

Oxygen leaves the furnace in the top gas only. In terms of variables already considered, its quantity may be described by:

$$n_O^o = n_C^g \cdot (O/C)^g \qquad (4.7)$$

where n_C^g is moles of carbon leaving in the top gas per mole of product Fe and $(O/C)^g$ is moles of oxygen leaving in the top gas per mole of carbon leaving in the top gas (Section 4.1.2).

4.2 The Stoichiometric Equation

Section 4.1 has presented seven independent equations, (4.1) to (4.7), in terms of eleven variables, n_{Fe}^i, n_{Fe}^o, n_C^i, n_C^o, n_O^i, n_O^o, n_C^g, $(C/Fe)^m$, n_O^B, $(O/Fe)^x$ and $(O/C)^g$.

This means that four variables must be specified before the steady-state operation of a blast furnace is fully described. Of the above variables, we can normally expect ore composition $(O/Fe)^x$ and pig-iron composition $(C/Fe)^m$ to be specified before we attempt to define or predict how a particular blast furnace must operate. This leaves two more variables to be specified or two more equations to be developed before an *a priori* model of the blast furnace is complete.

The stoichiometric equations in Section 4.1 can be consolidated by combining equations (4.3), (4.6) and (4.7) to give:

$$n_O^B + (O/Fe)^x = n_C^g \cdot (O/C)^g \qquad (4.8)$$

where $(O/Fe)^x$ and $(O/C)^g$ are the compositions of the incoming iron oxides and the outgoing gases. Once $(O/Fe)^x$ is specified, we have one equation and three unknown variables, which is consistent with the need to find two more specifications to complete the mathematical description.

4.2.1 Specific Uses for Carbon

It is convenient at this point to divide the functions of the input carbon into (i) the moles of active carbon, n_C^A, which react with oxygen in the blast and iron oxides; and (ii) the moles of inactive carbon which simply dissolve in the molten pig iron, $(C/Fe)^m$, both per mole of product Fe. Of course, all the active carbon ends up as CO or CO_2 in the top gas so that:

$$n_C^A = n_C^g.$$

In these terms,* equation (4.8) becomes

$$n_O^B + (O/Fe)^x = n_C^A \cdot (O/C)^g. \tag{4}$$

Equation (4) is the stoichiometric equation for the blast furnace in its most direct and simplest form. It is the end result of this chapter and it is referred to throughout the remainder of the text.

4.3 Calculations

Although not yet in a particularly meaningful form, equation (4) can be usefully employed. It can, for example, be used to check the consistency of measured top-gas compositions with the type of ore and measured inputs of:

(a) carbon, per tonne of molten pig iron;
(b) blast oxygen, per tonne of molten pig iron.

These items can be readily obtained from daily production sheets.

Consider, for example, calculation of the top gas composition for a blast-furnace operation similar to that described in Fig. 1.3, the data for which are shown in Table 4.1.

Insertion of these data into equation (4) leads to:

$$1.22 + 3/2 = 1.85 \cdot (O/C)^g$$

from which

$$(O/C)^g = 1.47.$$

---

*Similarly, when coke is the only source of carbon, the input/output carbon balance $(n_C^i = n_C^o)$ becomes:

$$n_C^{coke} = n_C^A + (C/Fe)^m.$$

Blast-furnace iron (inside the furnace) is saturated with carbon, 5 wt.% C $[(C/Fe)^m \cong 0.25]$ at 1800 K. $(C/Fe)^m$ is assigned with the value of 0.25 throughout the text.

Blast-Furnace Stoichiometry

TABLE 4.1

Item	Specification	Quantity kg per tonne of Fe	Quantity kg moles per tonne of Fe	Model variable (kg moles per kg mole of product Fe)
Fe		1000	17.9	
Iron oxide	Fe_2O_3			$(O/Fe)^x = 3/2$
Pig iron	5% C	53	4.4	$(C/Fe)^m = 0.25$
		inactive carbon		
Coke (only carbon source)	90% C	500 (450 kg of carbon)	37.5 (carbon)	$n_C^i = 2.10$
Active carbon		397	33.1	$n_C^A = n_C^i - (C/Fe)^m = 1.85$
Oxygen from blast air		350	21.9 (O)	$n_O^B = 1.22$

This O/C ratio is equivalent (Section 4.1.2) to CO and CO_2 mole fractions of $X_{CO_2}^g = 0.47$ and $X_{CO}^g = 0.53$. Furthermore, since

$$n_{CO}^g = n_C^A \cdot X_{CO}^g = 0.98 \text{ moles per mole of product Fe;}$$

$$n_{CO_2}^g = n_C^A \cdot X_{CO_2}^g = 0.87 \text{ moles per mole of product Fe;}$$

$$n_{N_2}^g = \frac{0.79}{0.21} \cdot 1/2 \cdot n_O^B = 2.29 \text{ moles per mole of product Fe;}$$

the overall top gas composition is 24 vol.% CO, 21% CO_2 and 55% N_2.

Several problems illustrating the use of equation (4) are included at the end of this chapter.

4.4 Graphical Representation of the Stoichiometric Balance

For graphical purposes, equation (4) may be rewritten:

$$(O/Fe)^x - (-n_O^B) = n_C^A \cdot \{(O/C)^g - \underbrace{0}_{zero}\}. \tag{4a}$$

$(O/Fe)^x$ and n_O^B are both expressed in terms of moles of oxygen (O) per mole of product Fe, hence equation (4a) has the form:

$$y_2 - y_1 = M \cdot \{x_2 - x_1\} \tag{4.10}$$

which is a straight line of slope M passing through the points x_1, y_1 and x_2, y_2.

Equation (4) can, therefore, be plotted on a graph with axes O/Fe and O/C to give a straight line of slope n_C^A joining the points

$$O/C = 0, \qquad O/Fe = -n_O^B$$
$$O/C = (O/C)^g, \; O/Fe = (O/Fe)^x$$

as is shown in Fig. 4.2.

This is the formal basis for the 'Rist Diagram' as it will be used in the remainder of this text (Rist, 1977). As will be seen later, the diagram can be used in conjunction with enthalpy and approach-to-equilibrium data to obtain a graphical view of (i) the efficiency with which a furnace is operating and (ii) the effects of changes in operating parameters. The diagram is a useful adjunct to the analytical equations of the model but it is not an indispensable part of their development.

4.4.1 Graphical Calculations

The stoichiometric diagram can be used directly for the type of calculation solved numerically in Section 4.3.

For example, consider the somewhat inefficient blast-furnace operation described in Table 4.2.

The top-gas composition for this furnace can be calculated by plotting:

$$O/C = 0, \quad O/Fe = -n_O^B = -1.40$$

and by drawing a straight line of slope

$$n_C^A = 2.05$$

to the point

$$O/C = (O/C)^g = ?, \quad O/Fe = (O/Fe)^x = 3/2.$$

This has been done on Fig. 4.2 from which $(O/C)^g = 1.41$, which is equivalent to an analysis of 41% CO_2 and 59% CO in the carbonaceous portion of the top gas.

High coke, high blast operations of this type with a high proportion of CO in the top gas are typical of small (1000 tonnes per day), old blast furnaces.

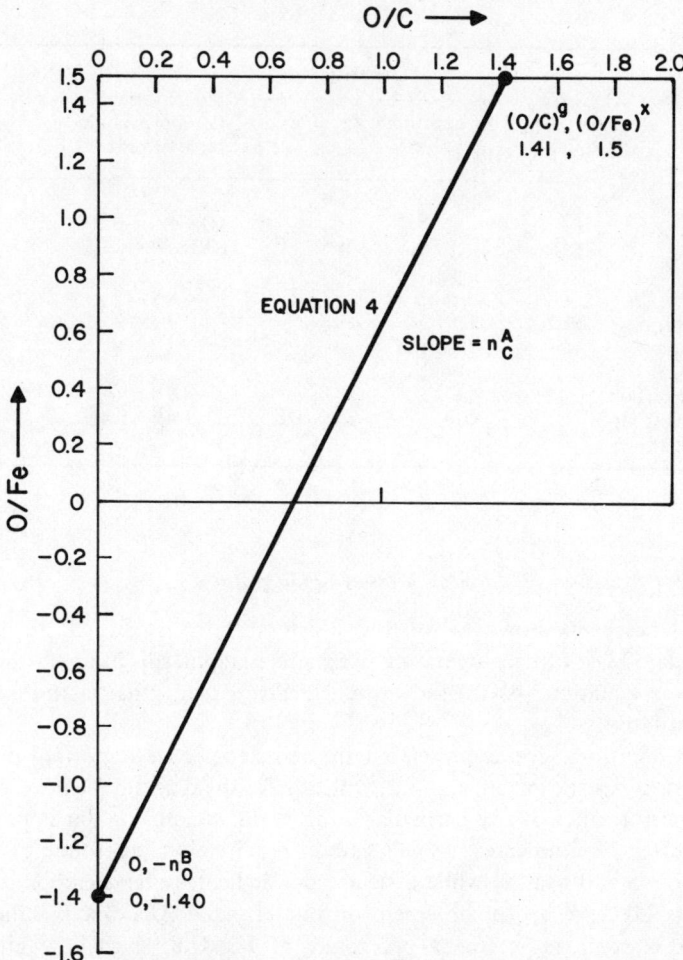

Fig. 4.2. Graphical representation of equation (4), the blast-furnace stoichiometric equation. The numerical values of n_O^B, n_C^A and $(O/C)^g$ are for the blast-furnace operation described in Table 4.2. Physical meanings of the slope and intercepts are described in Fig. 4.3.

TABLE 4.2

Item	Specification	Quantity kg per tonne of Fe	Quantity kg moles per tonne of Fe	Model variable (kg moles per kg mole of product Fe)
Fe		1000	17.9	
Iron oxide	Fe_2O_3			$(O/Fe)^x = 3/2$
Pig iron	5% C	53 inactive carbon	4.4	$(C/Fe)^m = 0.25$
Coke (only carbon source)	90% C	550 (495 kg of carbon)	41.3	$n_C^i = 2.3$
Active carbon		442	36.8	$n_C^A = n_C^i - (C/Fe)^m = 2.05$
Oxygen from blast air		400	25.0 (O)	$n_O^B = 1.40$

4.4.2 Discussion of the Stoichiometric Diagram

As has been shown above, the stoichiometric diagram can be used to calculate blast-furnace-operating parameters graphically in much the same way as equation (4) is used arithmetically. All lengths on the axes are quantitative.

In addition, several points, lengths and slopes have important physical meanings, as shown on Fig. 4.3, which make the diagram ideal for judging the performance of any particular furnace. For example, a flat slope of the *operating line* indicates a small carbon requirement per tonne of Fe for reduction and heating while a steep slope indicates a large carbon requirement. Likewise a deep intercept on the left-hand (O/Fe) axis indicates a large blast-air requirement per tonne of Fe while a shallow intercept indicates the opposite.

As will be shown in Chapters 6 and 9, the actual position and slope of the *operating line* depend upon enthalpy considerations in the furnace and the extent to which the reduction processes approach equilibrium. Variations in these conditions between blast furnaces are also readily visualized on the diagram.

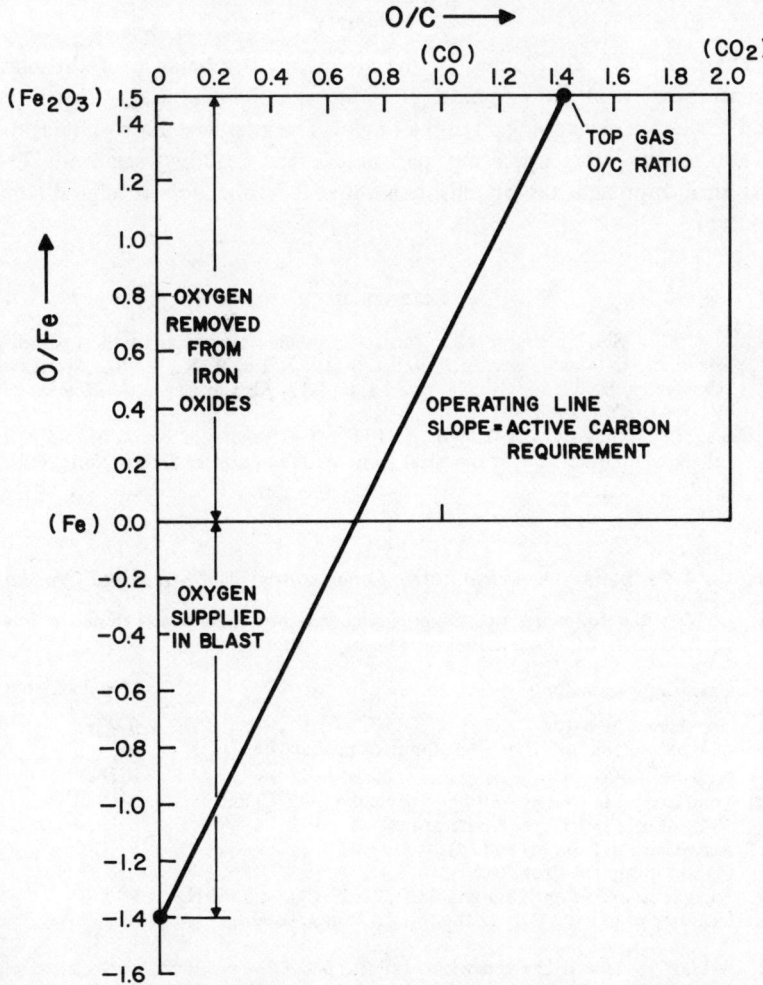

Fig. 4.3. Blast-furnace-operating diagram showing the *operating line* defined by equation (4). The physical meanings of its slope and several points and lengths are also indicated. All distances are quantitative.

4.5 Summary

There are two approaches to the stoichiometric balance of the blast furnace: analytical and graphical. Both are equally valid and both will be used further in developing a unified model. The graphical approach has the advantage that the operating parameters are readily visualized. The analytical approach has the advantage that it is amenable to digital computation.

References

Rist, A. (1977) 'Solving simple blast furnace problems by means of the operating diagram', in *Blast Furnace Ironmaking 1977*, Lu, W. K., Editor, McMaster University, Hamilton, Canada, pp. 4-1 to 4-37. Also in *Revue de Métallurgie*, Vols. 61 to 63, 1964 to 1966.

Tsuchiya, N., Tokuda, M. and Ohtani, M. (1976) 'The transfer of silicon from the gas phase to molten iron in the blast furnace', *Metallurgical Transactions. B*, 7B, 316.

Chapter 4 Problems. *Stoichiometric Calculations: Analytical and Graphical*

4.1 Convert the following operating parameters into numerical values for their equivalent blast furnace model variables.

Operating parameter	Model variable
(a) Ore charge: hematite	$(O/Fe)^x$
(b) Blast: 1200 Nm³ of dry air per tonne of product Fe	n_O^B
(c) Pig iron: 4.9% C	$(C/Fe)^m$
(d) Total carbon in charge: 460 kg of dry coke (90% C) and 50 kg of oil (85% C) per tonne of Fe	$n_C^t = n_C^o$
(e) Active carbon from (c) and (d)	n_C^A
(f) Carbon in top gas (from (e))	n_C^g
(g) Top gas composition: 23.9 vol.% CO, 20.5% CO_2, 55.6% N_2.	$(O/C)^g$
(h) Quantity of CO and CO_2 in the top gas from (f) and (g)	$n_{CO}^g, n_{CO_2}^g$

4.2 A blast furnace is charged with hematite pellets. It produces a 5% carbon pig iron. 450 kg of carbon (as coke) and 370 kg of oxygen (as air blast) are supplied to the furnace per tonne of product Fe. Calculate analytically:

(a) the composition (including N_2) of the top gas from this furnace;
(b) the volume of top gas per kg mole and per tonne of product Fe.

4.3 Plot the 'OPERATING LINE' for the operation in Problem 4.2 on an O/C: O/Fe graph and check graphically your answer to part (a) of Problem 4.2. Express your comparative answers in terms of the oxygen/carbon ratio in the top gas.

Blast-Furnace Stoichiometry

4.4 The Problem 4.2 blast furnace is switched to an acid, low SiO_2 sinter in which the predominant iron oxide is Fe_3O_4.* The low SiO_2 content of this feed leads to a low slag production (per tonne of product Fe) which permits the operator to cut back his coke and blast supplies to 425 kg of C and 345 kg of O_2 (per tonne of Fe) respectively. Demonstrate graphically the effects of this change and determine the mole fractions of CO and CO_2 in the carbonaceous portion of the top gas (X_{CO}, X_{CO_2}) for this new operation.

4.5 The operator of a small, old blast furnace wishes to check his blast-metering devices on the basis of top-gas analysis and coke-rate measurements. Determine, analytically or graphically, what his total blast volume (Nm^3 of dry air) should have been for the following operating conditions:

Item	Quantity
Ore: hematite (5% SiO_2, all of which goes to slag)	
Hot metal (5% C)	1000 tonnes
Coke (88% C, dry basis)	700 tonnes
	(a poor operation)
Top-gas composition	vol.% CO/vol.% CO_2 = 1.2

4.6 The composition of the top gas from a hematite-charged furnace is 22 vol.% CO, 20 vol.% CO_2, 58 vol.% N_2. The blast volume is 1400 Nm^3 per tonne of product Fe. From these data, calculate graphically the quantity of carbon taking part in the blast-furnace reactions (per tonne of product Fe). Calculate also the total carbon being used by the furnace if the pig iron contains 5 wt.% C.

4.7 The equations of this chapter neglect minor reduction reactions. However, these reactions, especially reduction of silica, do affect the operating parameters of the blast furnace and they should be included.

Add a single term to equation (4) which will show the effects of silicon-in-iron on the operating parameters of the furnace.

4.8 Prepare a computer programme which will:
   (a) convert measured blast-furnace parameters into equation (4) variables n_O^B, $(O/Fe)^x$, n_C^A, $(O/C)^g$;
   (b) use any three of these variables (e.g. n_O^B, $(O/Fe)^x$, n_C^A) to calculate the fourth (e.g. $(O/C)^g$);
   (c) convert the outputs from (b) into measured furnace parmeters, including complete top-gas analysis and top-gas volume (per tonne of Fe).
Employ the programme to check your answers to Problems 4.2, 4.4, 4.5 and 4.6.

4.9 Prepare a computer programme which will plot equation (4) on O/C, O/Fe axes from the specifications given in Problem 4.8(b). The intercepts at O/C = 0, O/Fe = $(O/Fe)^x$ and the slope of the operating line should be labelled.

*Limons, R. A. (1977) 'Iron ore − sinter', in *Blast Furnace Ironmaking 1977*, Lu, W. K., Editor, McMaster University, Hamilton, Canada, pp. 14-23 to 14-28.

CHAPTER 5

Development of a Model Framework: Simplified Blast-Furnace Enthalpy Balance

To this point, the mathematical blast-furnace model consists of the equation:

$$n_O^B + (O/Fe)^x = n_C^A \cdot (O/C)^g \tag{4}$$

where:

n_O^B = moles of oxygen (O) in blast air per mole of Fe produced by the furnace;

n_C^A = moles of active carbon, i.e. carbon used for heating and reduction but excluding carbon dissolved in the pig iron product;

$(O/Fe)^x$ = composition of the incoming iron oxides;

$(O/C)^g$ = composition of the outgoing top gas.

As has been shown in Chapter 4, this equation is useful for checking the consistency of measured top-gas compositions with measured inputs to the furnace. It is not, however, powerful enough to predict, *a priori*, the quantities of coke and blast which are necessary for a given type of blast-furnace operation and furnace charge. This chapter moves towards such a fully predictive model by developing a second equation for the process, based upon a steady-state enthalpy balance for the furnace.

5.1 Simplifications for an Initial Enthalpy Balance

Figure 4.1 shows the simplifications which are made in the initial enthalpy balance. It is assumed that:

(a) the air blast enters the furnace at 298 K and the top gas leaves at

298 K, i.e. neither carries sensible heat, which permits the N_2 of the blast to be ignored for the moment;
(b) the only charge materials are Fe_2O_3 and C, i.e. there is no gangue, flux or coke ash and thus no slag;
(c) the enthalpy of the carbon in the pig iron product is negligible;
(d) there are no convective or radiative heat losses from the furnace.

These assumptions are made only to preserve simplicity in the initial approach and the effects of each are considered in detail later.

5.2 The Enthalpy Balance

The steady-state enthalpy balance for the blast furnace may be expressed in the form:

Enthalpy into the furnace per mole of product Fe = Enthalpy out of the furnace per mole of product Fe

or in algebraic terms:

$$n_{Fe_2O_3} \cdot H^\circ_{298} = H^\circ_{1800} + n^g_{CO} \cdot H^\circ_{298} + n^g_{CO_2} \cdot H^\circ_{298} \quad (5.1)$$
$$\phantom{n_{Fe_2O_3} \cdot H^\circ_{298} =} Fe_2O_3 \quad\; Fe_l \quad\;\; CO \quad\;\;\; CO_2$$

where $n_{Fe_2O_3}$, n^g_{CO}, $n^g_{CO_2}$ are moles of Fe_2O_3 charged to the furnace and moles of CO and CO_2 leaving the furnace all per mole of product Fe. The enthalpies of C, O_2 and N_2 are zero because they enter or leave the furnace at 298 K (Fig. 4.1).

Equation (5.1) may be conveniently rearranged to give positive numbers as follows:

$$n_{Fe_2O_3} \cdot (-H^\circ_{298}) + H^\circ_{1800} = n^g_{CO} \cdot (-H^\circ_{298}) + n^g_{CO_2} \cdot (-H^\circ_{298}).$$
$$\phantom{n_{Fe_2O_3} \cdot} Fe_2O_3 \quad\;\; Fe_l \quad\quad\quad\;\; CO \quad\quad\quad\;\; CO_2$$
$$(5.1a)$$

5.3 Heat Supply and Heat Demand

It is useful to consider the blast-furnace enthalpy balance in terms of an enthalpy demand and an enthalpy supply. In the present simplified case, for example, the demand is considered to be the enthalpy required to produce liquid Fe 1800 K from Fe_2O_3 at 298 K and the supply is the enthalpy provided by oxidation of carbon 298 K to CO and CO_2 at 298 K. The demand/supply form of the enthalpy equation is

$$\mathscr{D} = S.$$

From equation (5.1a) it can be seen that:

$$\mathscr{D} = n_{Fe_2O_3} \cdot (-H^o_{298\,Fe_2O_3}) + H^o_{1800\,Fe_l}$$

and

$$S = n^g_{CO} \cdot (-H^o_{298\,CO}) + n^g_{CO_2} \cdot (-H^o_{298\,CO_2}).$$

The advantages of this form of the enthalpy equation are:

(a) both sides of the equation contain positive numbers;
(b) an additional enthalpy requirement for the process, e.g. for heating and melting slag, can be included simply and clearly as an additional heat-demand term;
(c) an additional enthalpy supply, e.g. enthalpy in hot blast, can be incorporated as an additional supply term.

This demand/supply form also demonstrates that for a blast furnace to be operating at a steady state, its heat demand must be exactly met by its heat supply. Otherwise the furnace will be warming or cooling and an adjustment of coke and blast inputs will eventually have to be made.

5.3.1 Numerical Development

Numerical values which can be substituted into equation (5.1a) are (from Appendices V and VI):

$$H^o_{298\,Fe_2O_3} = H^f_{298\,Fe_2O_3} = -826\,000 \text{ kJ (kg mole of } Fe_2O_3)^{-1},$$

enthalpy of formation from elements, 298 K

$$H^o_{298\,CO} = H^f_{298\,CO} = -111\,000 \text{ kJ (kg mole of CO)}^{-1},$$

Simplified Enthalpy Balance

$$H^{\circ}_{298} = H^{f}_{298} = -394\,000 \text{ kJ (kg mole of } CO_2)^{-1},$$
$$\phantom{H^{\circ}_{298}}_{CO_2} _{CO_2}$$

$$H^{\circ}_{1800} = [H^{\circ}_{1800} - H^{\circ}_{298}] = 73\,000 \text{ kJ (kg mole of Fe)}^{-1},$$
$$\phantom{H^{\circ}_{1800}}_{Fe_l} _{Fe_l} _{Fe_s}$$

giving the numerical enthalpy balance:

$$n_{Fe_2O_3} \cdot 826\,000 + 73\,000 = n^{g}_{CO} \cdot 111\,000 + n^{g}_{CO_2} \cdot 394\,000. \quad (5.2).$$

5.3.2 Adaptation to Model Variables

It is useful to express the enthalpy equation in the same terms as equation (4) and this can be done by making the substitutions:

$$n^{g}_{CO} = n^{g}_{C} \cdot X^{g}_{CO} = n^{A}_{C} \cdot \{2 - (O/C)^g\} \quad \text{(Sections 4.1.2}$$
$$n^{g}_{CO_2} = n^{g}_{C} \cdot X^{g}_{CO_2} = n^{A}_{C} \cdot \{(O/C)^g - 1\} \quad \text{and 4.2.1)}$$

and by noting that one mole of product Fe requires $\frac{1}{2}$ mole of input Fe_2O_3, i.e.

$$n_{Fe_2O_3} = \tfrac{1}{2}.$$

With these substitutions, equation (5.2) becomes:

$$\tfrac{1}{2} \cdot 826\,000 + 73\,000 = n^{A}_{C} \cdot [\{2 - (O/C)^g\} \cdot 111\,000$$
$$+ \{(O/C)^g - 1\} \cdot 394\,000] \quad (5.3)$$

which finally simplifies to:

$$\mathscr{D} = S,$$
$$\tfrac{1}{2} \cdot 826\,000 + 73\,000 = n^{A}_{C} \cdot \{283\,000 \cdot (O/C)^g - 172\,000\}. \quad (5.4)$$

It will be noted that both sides of equation (5.4) are in terms of kJ per kg mole of product Fe.

5.4 A General Enthalpy Framework

The demand/supply approach outlined above is followed throughout the remainder of the text. Additional enthalpy demands and enthalpy supply terms are incorporated as they are encountered but the general framework of equations (5.3) and (5.4) is maintained throughout the chapters.

The only generalization which is useful at this point is to leave the demand term as a variable, in which case equation (5.4) becomes

$$\mathscr{D} = S.$$
$$= n_C^A \cdot \{283\,000 \cdot (O/C)^g - 172\,000\} \tag{5.5}$$

where $\mathscr{D}$ is the heat demand of the process per kg mole of Fe produced by the furnace.

This generalized $\mathscr{D}$ term facilitates enthalpy calculations for different iron oxides and different molten metal temperatures and, as is shown in the problems following this chapter, additional enthalpy demands are readily incorporated into it.

5.4.1 Two Examples of $\mathscr{D}$, the Heat Demand

As shown above, the heat demand for producing one mole of Fe_l 1800 K from $\frac{1}{2}$ mole of Fe_2O_3 298 K is

$$\mathscr{D} = \tfrac{1}{2} \cdot (-H^f_{298}) + [H^\circ_{1800} - H^\circ_{298}]$$
$$\phantom{\mathscr{D} = \tfrac{1}{2} \cdot (}\;\;Fe_2O_3 \quad\quad\;\; Fe_l \quad\;\; Fe_s$$
$$= \tfrac{1}{2} \cdot 826\,000 + 73\,000 = 486\,000 \text{ kJ (kg mole of product Fe)}^{-1}.$$

Likewise the heat demand for producing one mole of Fe_l at 1900 K from Fe_3O_4 at 298 K is

$$\mathscr{D} = \tfrac{1}{3} \cdot (-H^f_{298}) + [H^\circ_{1900} - H^\circ_{298}]$$
$$\phantom{\mathscr{D} = \tfrac{1}{3} \cdot (}\;\;Fe_3O_4 \quad\quad\;\; Fe_l \quad\;\; Fe_s$$
$$= \tfrac{1}{3} \cdot (1\,121\,000) + 78\,000 = 452\,000 \text{ kJ (kg mole of Fe)}^{-1}.$$

It will be noticed that the heat demand $\mathscr{D}$ is calculated independently of the variables n_C^A and $(O/C)^g$, i.e. its value depends only upon the nature of the charge, the composition and temperature of the iron and slag and heat losses. Of course a change in $\mathscr{D}$, due say to an increased slag production per tonne of product Fe, will necessitate changes in n_C^A and $(O/C)^g$ for the furnace to re-establish steady-state operating conditions.

5.5 Summary

A second equation has been developed for the blast furnace based upon an overall steady-state enthalpy balance for the process. This equation can now be combined with the Chapter 4 stoichiometric equation to more completely define furnace-operating requirements.

In addition, it has been shown that the blast-furnace enthalpy balance can be divided into reactions which supply heat to the process, e.g. formation of CO and CO_2, and reactions which require heat, e.g. dissociation of iron oxides and heating and melting of product iron. This concept is useful when considering the many other factors which contribute to the total enthalpy balance.

Chapter 5 Problems. *Heat Demand and Heat Supply*

5.1 Calculate blast-furnace heat demands (kJ per kg mole of Fe) for the following cases:

(a) Pure Fe_2O_3 ore, pure carbon charge (298 K); pure liquid iron product (1800 K).

(b) Case (a) with an improvement in accuracy by taking into account that the iron contains 5% carbon. Neglect the heat of mixing.

(c) Case (b) with an improvement in accuracy by taking into account manganese (charged as MnO_2) and silicon in the iron (1% each). Neglect heats of mixing.

(d) Case (c) plus convective and radiative heat losses. Convective and radiative heat losses from large blast furnaces are in the order of:

$$8 \times 10^6 \times \text{(hearth diameter, m)} \qquad \text{(kJ hr}^{-1}\text{)}$$

Use the Burns Harbour 'D' furnace example (Table 1.1) for other data (assume that the product metal is 93% Fe).

(e) Case (d) plus the heat required for slag heating and melting. Assume that: (i) the ore contains 7% SiO_2, (ii) the CaO/SiO_2 weight ratio of the slag is 1.2 and (iii) CaO is charged as CaO in sinter rather than as $CaCO_3$. Neglect for now, coke ash and other slag components. All MnO_2 charged to the furnace is reduced to Mn.

Data:

$[H°_{1800} - H°_{298}]_{CaO} = 79\,000$ kJ (kg mole)$^{-1}$,

$[H°_{1800} - H°_{298}]_{SiO_2} = 105\,000$ kJ (kg mole)$^{-1}$,

H^f_{1800} (from CaO and SiO_2) $= -400$ kJ (kg of slag)$^{-1}$.
   liquid slag

5.2 The CO/CO_2 ratio of the top gas leaving a hematite-charged blast-furnace is approximately 1. The carbon supply rate (including carbon-in-iron, 5 wt% C) is 500 kg per tonne of product Fe. What is the enthalpy supply to this furnace (kJ per kg mole of product Fe), assuming that the blast enters and the top gas leaves at 298 K? How can this enthalpy supply be increased without increasing the fuel or oxygen supplies?

CHAPTER 6

The Model Framework: Combinations of Stoichiometric and Enthalpy Equations

This chapter combines the stoichiometric equation derived in Chapter 4 and the enthalpy balance equation derived in Chapter 5 to illustrate further the methods by which a final mathematical description of the blast furnace is developed. The two equations are:*

$$n_O^B + (O/Fe)^x = n_C^A \cdot (O/C)^g, \qquad (4)$$

$$\mathscr{D} = \underset{\text{heat supply of the process}}{S}$$

$$= n_C^A \cdot [283\,000 \cdot (O/C)^g - 172\,000], \qquad (5.5)$$

where $\mathscr{D}$ is the heat demand for dissociating the input iron oxide and for heating the iron product.

To this point, then, the mathematical model consists of two equations and five unknowns which indicates that even under the highly simplified conditions of Chapters 4 and 5, three operating parameters must be specified to fully define a blast-furnace operation.

Ore composition $(O/Fe)^x$ is one such specification and a second comes from the fact that there is a specific relationship between a given ore and the value of $\mathscr{D}$, its heat demand (Section 5.4). Unfortunately, the third and final specification cannot be obtained by considering the blast furnace

---

*Where, it will be remembered,

$$283\,000 = H^f_{298,\,CO} - H^f_{298,\,CO_2} \text{ and } 172\,000 = 2H^f_{298,\,CO} - H^f_{298,\,CO_2} \qquad \text{(Section 5.3)}$$

as a whole. It is obtained (Chapters 7, 8 and 9) by treating the furnace as two segments, separated by a conceptual division through its chemical reserve zone.

6.1 Combining Stoichiometric and Enthalpy Equations: Calculations

Equation (5.5) is readily rearranged to

$$\frac{\mathcal{D}}{283\,000} = n_C^A \cdot (O/C)^g - n_C^A \cdot \frac{172\,000}{283\,000}. \tag{5.5a}$$

Subtraction of this form of equation (5.5) from equation (4) leads to:

$$n_O^B + (O/Fe)^x - \frac{\mathcal{D}}{283\,000} = n_C^A \cdot \frac{172\,000}{283\,000} \tag{6.1}$$

which shows that once $(O/Fe)^x$ and $\mathcal{D}$ are specified (i.e. from the nature of the ore and its heat demand) the model consists of one equation and two unknowns. This means that specification of either n_C^A (from the input carbon per mole of product Fe and the carbon content of the pig iron) or n_B^O (oxygen in dry blast air per mole of product Fe) fully defines the Fig. 4.1 blast-furnace operation.

6.1.1 Calculations

Equation (6.1), though restricted to the conditions described in Chapter 4 (Fig. 4.1), can be used to give a preliminary estimate of the quantity of carbon which will satisfy a particular amount of air supply to the furnace (or vice versa). It might be used, for example, to indicate the rate at which carbon must be supplied to a furnace for any given wind rate.

As an example calculation, consider the use of equation (6.1) to estimate the coke requirement and top-gas analysis of a blast furnace which is operating according to Table 6.1.

Substitution of n_O^B, $(O/Fe)^x$ and $\mathcal{D}$ from Table 6.1 into equation (6.1) leads to

$$1.41 + 3/2 - \frac{486\,000}{283\,000} = n_C^A \cdot \frac{172\,000}{283\,000}$$

TABLE 6.1

Item	Specification	Quantity		Model Variable (kg moles per kg mole of product Fe)
		kg per tonne of Fe	kg moles per tonne of Fe	
Fe		1000	17.9	
Iron oxide	Fe_2O_3			$(O/Fe)^x = 3/2$
Oxygen from blast air	1350 Nm^3 of air per tonne of Fe	405 (oxygen)	25.3 (O)	$n_O^B = 1.41$
Heat demand				$\mathscr{D}$ = 486 000 kJ per kg mole of product Fe (Section 5.4.1)
Pig iron	5% C			$(C/Fe)^m = 0.25$
Active carbon				$n_C^A = ?$
Input carbon from all sources				$n_C^i = (C/Fe)^m + n_C^A$

from which
$$n_C^A = 1.96$$
and
$$n_C^i = \text{total input carbon} = (C/Fe)^m + n_C^A$$
$$= 0.25 + 1.96 = 2.21.$$
This value of n_C^i is equivalent to 39.6 kg moles or 475 kg of input carbon per tonne of Fe.

6.1.2 Gas Composition Calculation

The relative proportions of CO and CO_2 in the top gas can now be calculated from n_O^B, $(O/Fe)^x$ and n_C^A using equation (4) as is demonstrated in Section 4.3. For the operation described above, the O/C ratio of the top gas is 1.48, and the carbonaceous portion of the gas will contain 52% CO and 48% CO_2.

6.2 Graphical Representation of the Combined Stoichiometric–Enthalpy Equation

An alternative arrangement of equation (6.1) is:
$$\left\{(O/Fe)^x - \frac{\mathscr{D}}{283\,000}\right\} - (-n_O^B) = n_C^A \cdot \left\{\frac{172\,000}{283\,000} - \underset{(zero)}{0}\right\}$$

This equation has the form:
$$\{y_2\} - (y_1) = M \cdot \{x_2 - x_1\}$$
(i.e. it describes a line of slope M between points x_1; y_1 and x_2; y_2) and thus it may be plotted on O/C, O/Fe axes in much the same way as equation (4) (Section 4.4). On these axes it describes a straight line of slope n_C^A passing through the points:

O/C = 0,   O/Fe = $-n_O^B$

$O/C = \dfrac{172\,000}{283\,000}$,   O/Fe = $(O/Fe)^x - \dfrac{\mathscr{D}}{283\,000}$

as is plotted in Fig. 6.1.

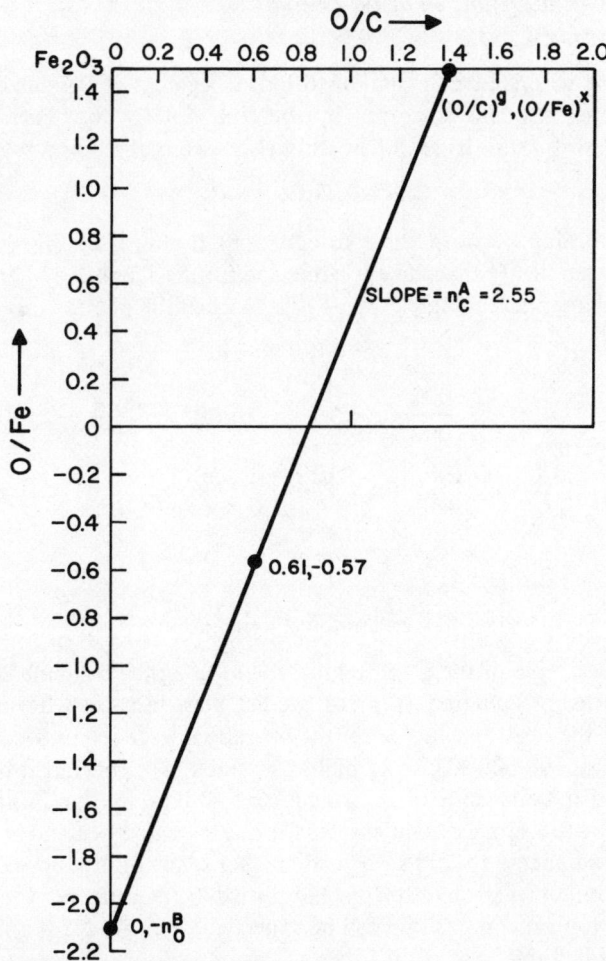

Fig. 6.1. Graphical representation of equations (6.1) and (4). The numerical values at the mid-point represent the coordinates:

$$O/C = \frac{172\,000}{283\,000}, \qquad O/Fe = (O/Fe)^x - \frac{\mathscr{D}}{283\,000}$$

for the Section 6.3 calculation. The slope and the intercepts are also from that calculation.

6.2.1 Graphical Definition of the Overall Operating Line

The line described by equation (6.1) is a segment of the line described by equation (4). This is proven by the fact that (i) both lines have the same slope (n_C^A) and (ii) both pass through the common point:

$$O/C = 0, \quad O/Fe = -n_O^B.$$

The relationship between the two equations is shown in Fig. 6.1 which demonstrates clearly that the operating line for the simplified blast-furnace operation described in Fig. 4.1 must pass through the points:

$$O/C = 0, \qquad O/Fe = -n_O^B,$$

$$O/C = \frac{172\,000}{283\,000}, \qquad O/Fe = (O/Fe)^x - \frac{\mathscr{D}}{283\,000},$$

$$O/C = (O/C)^g, \qquad O/Fe = (O/Fe)^x.$$
$$\text{top gas} \qquad\qquad \text{incoming iron oxide}$$

6.3 A Graphical Calculation

As an example of the graphical method of using the combined stoichiometric/enthalpy equation (6.1) to predict blast-furnace behaviour, consider that the heat demand, $\mathscr{D}$, of the operation described in Section 6.1.1 increases by 100 000 kJ per kg mole of product Fe. This might be caused by an exorbitant amount of gangue and fluxes in the furnace. The operator of the furnace compensates for this increased demand by increasing his coke charge to 600 kg of carbon (per tonne of product Fe) and he wants to know what quantity of blast air will be required per tonne of Fe and what top-gas composition can be expected. The data for this operation are shown in Table 6.2.

These data furnish directly the point:

$$O/C = \frac{172\,000}{283\,000} = 0.61; \qquad O/Fe = (O/Fe)^x - \frac{\mathscr{D}}{283\,000}$$

$$= 3/2 - \frac{586\,000}{283\,000} = -0.57$$

as is plotted in Fig. 6.1; and the slope of the operating line $n_C^A = 2.55$.

TABLE 6.2

Item	Specification	Quantity		Model variable (kg moles per kg mole of product Fe)
		kg per tonne of Fe	kg moles per tonne of Fe	
Fe		1000	17.9	
Iron oxide	Fe_2O_3			$(O/Fe)^x = 3/2$
Heat demand				$\mathscr{D}$ = 586 000 kJ per kg mole of product Fe
Pig iron	5% C			$(C/Fe)^m = 0.25$
Carbon charge (total)	Coke	600 (kg of carbon)	50.0	$n_C^i = 2.79$
Active carbon		547	45.6	$n_C^A = n_C^i - (C/Fe)^m = 2.55$
Oxygen from blast air		?		$n_O^B = ?$

72   The Iron Blast Furnace

The first step in calculating the blast-air requirement is to determine n_O^B, moles of O from blast air per mole of product Fe. This is done by (i) drawing a line of slope $n_C^A = 2.55$ through the above-mentioned point and by (ii) extending the line to the point:

$$O/C = 0; \qquad O/Fe = -n_O^B = ?$$

This calculation leads to $n_O^B = 2.12$ which is equivalent to 606 kg of oxygen or 2020 Nm³ of air per tonne of Fe.

Likewise, the composition of the top gas is obtained by extending the line of slope $n_C^A = 2.55$ to the point:

$$O/C = (O/C)^g = ? \qquad O/Fe = (O/Fe)^x = 3/2.$$

From this point $(O/C)^g = 1.42$, which is equivalent to a gas composition (carbonaceous portion) of 42% CO_2 and 58% CO.

This example demonstrates clearly the ease with which the graphical method can be used.

6.4 Summary and Discussion of Stoichiometry/Enthalpy Graph

The combined stoichiometric/enthalpy equation (6.1) developed in this chapter defines the combinations of carbon and blast air which will satisfy the enthalpy and stoichiometric requirements of the Fig. 4.1 blast furnace. It cannot, however, be used to specify the individual quantities of each.

In graphical terms the combined equation further defines the *Operating Line* (Section 4.5) of the furnace and shows that it must pass through the point

$$O/C = \frac{172\,000}{283\,000}; \qquad O/Fe = (O/Fe)^x - \frac{\mathscr{D}}{283\,000}$$

where $\mathscr{D}$ is the heat demand of the particular furnace charge. This specification and the fact that top-gas compositions must be between $(O/C)^g = 1$ and $(O/C)^g = 2$ (i.e. between pure CO and pure CO_2) show clearly that the Fig. 4.1 furnace must operate within the shaded wedges of Fig. 6.2.

The task of the remainder of the text is to locate precisely the position of the operating line within these wedges, i.e. to uniquely define the

The Model Framework

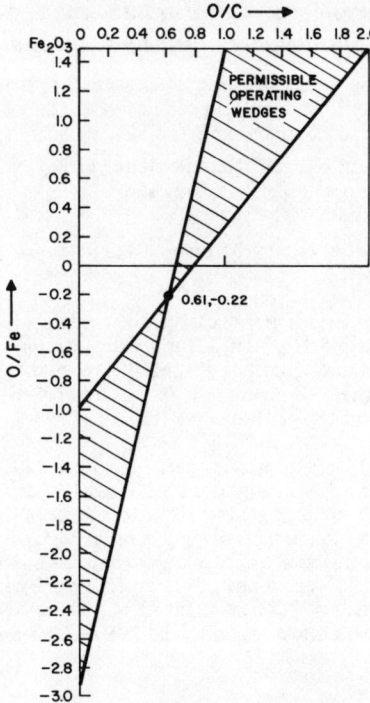

Fig. 6.2. Blast-furnace-operating diagram showing the typical 'wedges' of permissible *operating line* positions as fixed by stoichiometric and enthalpy considerations. The numerical values at the mid-point are for the specific case of a hematite charge $[(O/Fe)^x = 3/2]$ and a heat demand, $\mathscr{D}$, of 486 000 kJ per kg mole of Fe.

individual coke and blast requirements of the process. This requires that:

(a) the equations be improved to include blast temperature as an independent variable and top gas temperature as a dependent variable;
(b) that an additional operating equation be developed.

Suggested Reading

Rist, A. and Meysson, N. (1966) 'A dual graphic representation of the blast-furnace mass and heat balances', in *Ironmaking Proceedings, Philadelphia* (AIME), 25, 88–98.
(Of particular interest is the Discussion and Authors' Reply at the end of this paper.)

74   *The Iron Blast Furnace*

Chapter 6 Problems. *Calculations with the Combined Stoichiometric–Enthalpy Equation*

6.1 The heat demand of a hematite-charged blast furnace is 560 000 kJ per kg mole of product Fe. The total carbon in the charge is 600 kg per tonne of product Fe. The pig iron product contains 5% C. Calculate analytically:

(a) the volume of blast air (Nm^3 per tonne of Fe) which is required to keep this furnace operating at a steady state;
(b) the composition of the top gas.

(Assume that the blast enters the furnace at 298 K and that the top gas leaves at 298 K.)

6.2 A 7000-tonnes-per-day (of Fe) blast furnace is operating with the heat demand described in Problem 5.1(b). 8500 Nm^3 of air (assume 298 K) is blasted into the furnace per minute. Calculate graphically the rate at which carbon must be charged to the furnace to maintain it in balance thermally and stoichiometrically. Assume that the top gas leaves the furnace at 298 K.

6.3 A hematite-charged blast furnace is operating with a coke supply of 590 kg of C per tonne of Fe and a blast supply of 1850 Nm^3 of dry blast per tonne of Fe. The normal supply of high-grade pellets to the plant is interrupted and the operators are forced to charge a low-grade lump hematite ore to the furnace. It is found that with this new ore, the steady-state carbon and blast supplies are 650 kg and 2200 Nm^3 per tonne of Fe respectively. By how much has the new low-grade ore increased the heat demand of the process? Assume that blast enters the furnace and top gas leaves at 298 K. The product metal contains 5 wt% C.

CHAPTER 7

Completion of the Stoichiometric Part of the Model: Conceptual Division of the Blast Furnace through the Chemical Reserve Zone

Chapters 4, 5 and 6 demonstrated in greatly simplified terms the methods by which a fully predictive blast-furnace model can be developed. These chapters employed carbon, iron, oxygen and enthalpy balances over the entire furnace and these led to the operating equation:

$$n_O^B + (O/Fe)^x - \frac{\mathscr{D}}{283\,000} = n_C^A \cdot \frac{172\,000}{283\,000} \qquad (6.1)$$

where:

$\mathscr{D}$ = heat demand for reduction, heating and melting (kJ per kg mole of product Fe),

n_O^B = oxygen supplied in blast air (kg moles of O per kg mole of product Fe),

$(O/Fe)^x$ = molecular oxygen/iron ratio in the input iron oxides,

n_C^A = carbon taking part in heating and/or reduction reactions (kg moles of C per kg mole of product Fe),

$283\,000 = H^f_{298\,CO} - H^f_{298\,CO_2}$ (kJ per kg mole of O)

$172\,000 = 2H^f_{298\,CO} - H^f_{298\,CO_2}.$ (kJ per kg mole of C)

Equation (6.1) can be seen to be a single equation containing four operating variables. Two of these variables, the oxygen/iron ratio $(O/Fe)^x$

of the iron ore and its associated heat demand $\mathscr{D}$, will normally be specified in any blast furnace problem, but even with these specifications, equation (6.1) still contains two unknown variables (n_O^B, n_C^A). This means that the approach taken in Chapters 4, 5 and 6, illustrative as it is, can never lead to a fully predictive model.

The goal of a fully predictive model is attained in Chapters 7, 8 and 9 by:

(a) conceptually dividing the furnace, top from bottom, at some height within the chemical reserve zone;
(b) carrying out stoichiometric and enthalpy balances over the two segments.

As will become clear, the most important balances are those of the bottom segment.

7.1 The Blast Furnace as Two Separate Reactors

Visualization of the blast furnace as two reactors (i.e. top and bottom segments) is perfectly correct as long as the mass and enthalpy balance equations of each segment are consistent with continuity of mass and enthalpy transfer between the segments. For this to be true, the operating equations must be consistent with the following two conditions:

(a) The net mass of each element leaving the top segment across the division must equal the net mass of that element crossing into the bottom segment.
(b) The net enthalpy in the material leaving the top segment across the division must equal the net enthalpy crossing into the bottom segment.

If these continuity conditions are scrupulously adhered to while developing the stoichiometric and enthalpy equations of the segments, it follows that operating parameters calculated for either segment will satisfy (i) the operating equations of both segments and thus (ii) the operating equations of the whole furnace. Otherwise, the furnace would not be operating at a steady state throughout.

Conceptual division of the blast furnace specifically through its chemical reserve zone has the advantage that it imposes two new conditions

The Stoichiometric Part of the Model

which must be described by the enthalpy and stoichiometric equations of the segments. These arise from the observations (Section 3.5) that:

(c) there is no carbon gasification in or above the chemical reserve zone, i.e. all carbon in the charge descends through the top segment into the bottom before it takes part in any chemical reactions;

(d) the only iron-bearing material crossing the division is wustite, $Fe_{0.947}O$, i.e. all higher oxides have been reduced to $Fe_{0.947}O$ by the time they descend into the chemical reserve but all reduction to Fe takes place below the chemical reserve.

As can be seen from the above discussion, the remainder of the development assumes that iron blast furnaces always contain a stable chemical reserve zone. Evidence to support this assumption has been provided in Chapter 3 and it is examined again in Chapter 10 as part of an overall criticism of the equations. In the meantime the presence of a stable chemical reserve is taken as fact.

The position of the chemical reserve zone in the furnace is described in

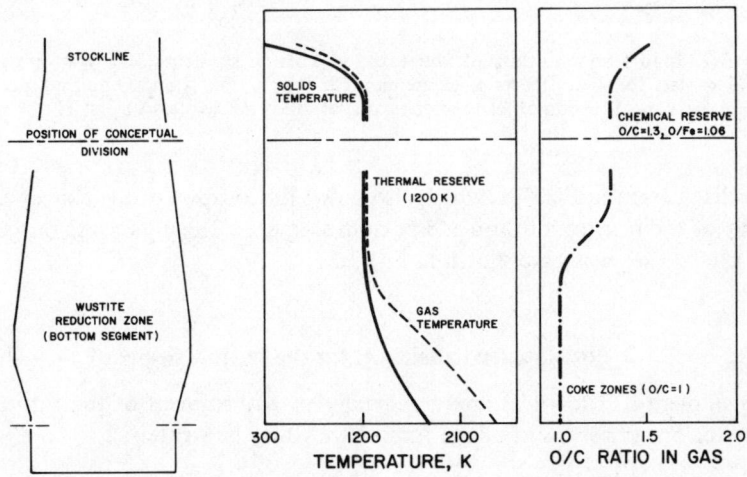

Fig. 7.1. Conceptual division of the blast furnace into top and bottom segments. The position of the dividing plane in relation to the chemical and thermal reserve zones is shown.

The Iron Blast Furnace

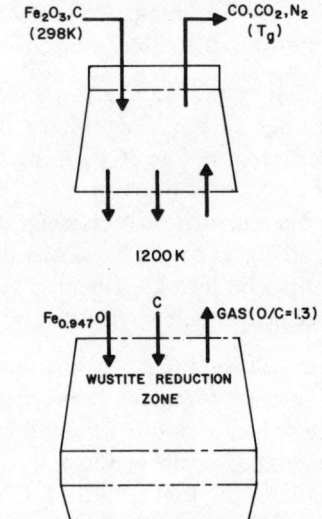

Fig. 7.2. Inputs and outputs of the top and bottom segments of a conceptually divided blast furnace. Effects of elements other than C, Fe, N and O and effects of sources of carbon other than coke are described in Chapters 11 and 12.

idealized form in Fig. 7.1, which also shows the location of the conceptual split of the furnace into top and bottom segments. The inputs and outputs of the two segments are shown in Fig. 7.2.

7.2 Stoichiometric Balances for the Bottom Segment

Equations (4.1) to (4.7) apply generally to any segment of the furnace. For the bottom segment of the furnace they may be written:

$$n_{Fe}^{iwrz} = n_{Fe}^{owrz} \qquad (4.1\ wrz)$$

$$n_{C}^{iwrz} = n_{C}^{owrz} \qquad (4.2\ wrz)$$

The Stoichiometric Part of the Model

$$n_O^{iwrz} = n_O^{owrz} \quad (4.3 \text{ wrz})$$

$$n_{Fe}^{owrz} = 1 \quad (4.4 \text{ wrz})$$

$$n_C^{owrz} = n_C^{gwrz} + (C/Fe)^m \quad (4.5 \text{ wrz})$$

$$n_O^{owrz} = n_C^{gwrz} \cdot (O/C)^{gwrz} \quad (4.6 \text{ wrz})$$

$$n_O^{iwrz} = n_O^B + (O/Fe)^{xwrz} \quad (4.7 \text{ wrz})$$

where $(C/Fe)^m$ and n_O^B are applicable to both the bottom segment and the whole furnace. The abbreviation wrz refers to the bottom segment of the furnace. Throughout the remainder of the text this segment is referred to as the wustite reduction zone because it is in this segment that final reduction occurs.

7.2.1 Bottom Segment Substitutions

As stated in Section 7.1, conditions in and above the chemical reserve zone are such that coke does not take part in any chemical reactions in the top segment of the furnace. This means that the carbon of the charge descends unreacted into the wustite reduction zone and that, as a consequence, the amount of carbon entering the wustite reduction zone is exactly the same as the amount entering the furnace,* i.e.

$$n_C^{iwrz} = n_C^i.$$

Similarly, because the amounts of carbon entering the furnace and the wustite reduction zone are identical, the amounts leaving must also be identical, i.e.

$$n_C^{owrz} = n_C^o.$$

This equation is important because it and equations (4.5) and (4.5 wrz) lead to:

$$n_C^{gwrz} + (C/Fe)^m = n_C^g + (C/Fe)^m$$

---

*The effects of sources of carbon other than coke (e.g. $CaCO_3$, hydrocarbons) are considered in Chapters 11 and 12.

or

$$n_C^{gwrz} = n_C^g.$$

This last equation is consistent with the assumption that in and above the chemical reserve zone there are no reactions with solid carbon to form CO or CO_2. This being the case, the amount of carbon ascending as CO and CO_2 from the bottom segment must, as the equation shows, be the same as the amount of carbon leaving the furnace in the top gas.

7.3 Stoichiometric Equation for the Wustite Reduction Zone

Following the procedures of Section 4.2, the stoichiometric equation (oxygen balance) for the wustite reduction zone is:

$$n_O^B + (O/Fe)^{xwrz} = n_C^{gwrz} \cdot (O/C)^{gwrz}$$

which, with the substitution

$$n_C^{gwrz} = n_C^g = n_C^A \qquad \text{(Sections 4.2.1, 7.2.1)}$$

becomes

$$n_O^B + (O/Fe)^{xwrz} = n_C^A \cdot (O/C)^{gwrz}. \qquad (7.1)$$

This equation is applicable to the wustite reduction zone in the same manner as equation (4) applies to the entire furnace.

7.3.1 Numerical Substitutions

The discussion in Section 7.1 indicated that a conceptual division of the blast furnace through its chemical reserve zone imposes the condition that the only iron-bearing material entering the bottom segment is $Fe_{0.947}O$, which means that $(O/Fe)^{xwrz}$ in equation (7.1) is given by:

$$(O/Fe)^{xwrz} = 1.056 \simeq 1.06.$$

The composition of the gases leaving the bottom segment is somewhat less

The Stoichiometric Part of the Model

certain because:

(a) it will vary with the extent to which equilibrium is approached;
(b) it will vary slightly with the thermal reserve temperature.

For now, however, it will be assumed that the gas in the chemical reserve, i.e. the gas leaving the bottom segment, has the equilibrium composition for the reaction:

$$CO + Fe_{0.947}O \rightleftharpoons 0.947\, Fe + CO_2 \qquad (1.3)$$

at a thermal reserve temperature of 1200 K. This equilibrium composition is

$$X_{CO} = 0.7^*, \quad X_{CO_2} = 0.3$$

which in model terms is equivalent to

$$(O/C)^{gwrz} = 1.3.$$

7.3.2 The Final Equation

Substitution of these numerical values into equation (7.1) leads to

$$n_O^B + 1.06 = 1.3 \cdot n_C^A \qquad (7)$$

which is a single equation in terms of the two operating parameters n_O^B and n_C^A, moles of blast oxygen (O) and moles of active carbon per mole of product Fe.

7.4 Discussion and Summary

Equation (7) describes the stoichiometric balance for the bottom segment of the furnace only. However, since continuity conditions were adhered to throughout its development, operating parameters calculated with it will also satisfy mass and enthalpy balances for the top segment and the whole furnace. Consequently, equation (7) can legitimately be used to predict the overall operating requirements of the blast-furnace process.

*Rounded from $X_{CO} = 0.694$ and $X_{CO_2} = 0.306$, Stull (1970).

It might also be noted here that maintenance of continuity demands that the sum of the stoichiometric equations of the segments must equal the stoichiometric equation of the whole furnace. This being the case, the stoichiometric equation for the top segment is equation (4) minus equation (7).

In summary, this chapter has shown that a conceptual division of the furnace through its chemical reserve zone greatly increases the power of the model, i.e. stoichiometry alone provides an equation which is as powerful as the stoichiometric/enthalpy equation developed in Chapters 4, 5 and 6. In fact only one further equation* is required before the operating requirements of the furnace (n_C^A, n_O^B) are fully defined. This additional equation comes from the enthalpy balance in the wustite reduction zone as is shown in Chapter 8.

Reference

Stull, D. R., Prophet, H. *et al.* (1970) *JANAF Thermochemical Tables*, 2nd edition, United States Department of Commerce, Document NSRDS-NBS 37, Washington, June 1971.

Chapter 7 Problems. *Wustite Reduction Zone Stoichiometry*

7.1 The 'coke rate' of a large modern blast furnace is 500 kg of coke (88% C, 12% ash) per tonne of product metal (95% Fe, 5% C). It is believed that the furnace is operating in such a way that (i) it has a 1200 K thermal reserve zone and (ii) that in this thermal reserve the gases approach equilibrium with $Fe_{0.947}O$ and Fe.

Calculate, using these chemical and thermal reserve assumptions, the amount of blast (per tonne of Fe) which must be supplied to keep the furnace operating at a steady state.

7.2 The charge to the furnace of Problem 7.1 is temporarily switched from hematite to magnetite without altering the amounts of impurities (per tonne of

---

*It might be thought at this point that equations (6.1) and (7) could be combined to uniquely define n_C^A and n_O^B. In fact, such a combination is not permissible because equation (6.1) assumes a top-gas temperature of 298 K which is not compatible with the thermal reserve temperature condition used in the development of equation (7).

The Stoichiometric Part of the Model

Fe) being charged to the furnace. What effect will this change to magnetite have on:

(a) the carbon and blast requirements of the process?
(b) the top gas composition?

(Assume that (i) metal and slag composition, (ii) convective and radiative heat losses are unchanged. Continue with the thermal and chemical reserve assumptions of Problem 7.1.)

7.3 The number 7 blast furnace at Algoma produced 5000 tonnes of metal (94% Fe, 5% C, 1% Si) over a 24-hour period. Over the same period it used 6×10^6 Nm³ of dry air blast and 2.4×10^6 kg of dry coke. As a check on coke analysis, estimate what % C in coke would be consistent with steady state operation of the furnace over this 24-hour period. Include the affect of silicon in iron.

(Assume a 1200-K thermal reserve and a close approach to $Fe_{0.947}O$/Fe equilibrium. SiO_2 is reduced below the chemical reserve zone.)

7.4 Prepare a computer programme which will:
   (a) convert measured blast-furnace parameters into n_O^B, $(O/Fe)^x$, $(Si/Fe)^m$, n_C^A, $(O/C)^g$, $(C/Fe)^m$.
   (b) calculate n_O^B from n_C^A and vice versa;
   (c) calculate $(O/C)^g$ from n_B^O, n_C^A and $(O/Fe)^x$;
   (d) convert the outputs from (b) and (c) into measured parameters, including complete top-gas analysis and top-gas volume (per tonne of Fe).

Use, as far as possible, the programme you have already prepared for Problem 4.8 but incorporate the effect of silicon in iron. Employ the new programme to check your answers to Problems 7.1 and 7.3.

CHAPTER 8

Enthalpy Balance for the Bottom Segment of the Furnace

Chapter 7 showed how the blast furnace can be divided into two segments from the viewpoint of mass and enthalpy balances. It showed also that maintenance of continuity between the segments while developing stoichiometric and enthalpy equations ensures that the operating parameters calculated for any segment are applicable to the whole furnace.

Dividing the furnace, top from bottom, at a height within the chemical reserve led to the steady-state mass balance equation:

$$n_O^B + 1.06 = 1.3 \cdot n_C^A \qquad (7)$$

which is the stoichiometric operating equation for the bottom segment of the furnace. This equation is applicable to all furnaces in which equilibrium conditions are approached (Section 7.3) within a chemical reserve zone. Departures from this condition are discussed after the model is complete (Section 10.6).

In this chapter, a steady-state enthalpy balance is carried out over the bottom segment and a second equation in terms of the variables n_O^B and n_C^A is developed. It will be shown (Chapter 9) that this enthalpy equation plus equation (7) define fully the operating requirements of the blast-furnace process.

8.1 Enthalpy Balance for the Bottom Segment

The items in the enthalpy balance of the wustite reduction zone are shown in Fig. 7.2. $Fe_{0.947}O$ and C enter at the thermal reserve temperature (1200 K); gas of the chemical reserve composition leaves at 1200 K,

Enthalpy Balance

and pig iron is produced at 1800 K. Blast enters at a temperature T_B depending upon the particular operation.

The steady-state enthalpy balance (enthalpy in = enthalpy out, per mole of product Fe) for the wustite reduction zone is

$$n_{Fe_{0.947}O} \cdot H^o_{1200} \atop Fe_{0.947}O + n_C^{iwrz} \cdot H^o_{1200} \atop C + n_O^B \cdot \tfrac{1}{2} \cdot H^o_T \atop O_2^B + n_N^B \cdot \tfrac{1}{2} \cdot H^o_T \atop N_2^B$$

$$= H^o_{1800} \atop Fe_l + (C/Fe)^m \cdot H_{1800} \atop [C] + n_{CO}^{gwrz} \cdot H^o_{1200} \atop CO$$

$$+ n_{CO_2}^{gwrz} \cdot H^o_{1200} \atop CO_2 + n_N^B \cdot \tfrac{1}{2} \cdot H^o_{1200} \atop N_2 . \qquad (5.1 \text{ wrz})$$

It will be noticed that equation (5.1 wrz) contains many more terms than its equivalent for the whole furnace (equation (5.1)). This is because C, N_2 and O_2 enter (or enter and leave) at temperatures other than 298 K. Several substitutions which can be made immediately are:

(a) $n_{Fe_{0.947}O}| = 1.056 \cong 1.06$,

i.e. 1.06 moles of $Fe_{0.947}O$ are necessary to produce 1 mole of Fe;

(b) $n_C^{iwrz} = n_C^{owrz} = n_C^A + (C/Fe)^m$ \hfill (Section 7.2.1),

i.e. all carbon in the charge descends to the wustite reduction zone before reacting.

(c) $n_{CO}^{gwrz} = n_C^{gwrz} \cdot X_{CO}^{gwrz} = n_C^A \cdot \left\{2 - (O/C)^{gwrz}\right\}$

\hfill (Sections 5.3 and 7.3),

$n_{CO_2}^{gwrz} = n_C^{gwrz} \cdot X_{CO_2}^{gwrz} = n_C^A \cdot \left\{(O/C)^{gwrz} - 1\right\}$.

(d) n_N^B, moles of N in blast air per mole of product Fe, is related to n_O^B by

$$n_N^B = \frac{0.79}{0.21} \cdot n_O^B.$$

n_O^B is oxygen introduced in dry blast air. Oxygen from other blast components, i.e. pure oxygen and H_2O, is treated separately, Chapter 11.

Making these substitutions, the enthalpy balance becomes

$$1.06 \cdot H°_{1200 \atop Fe_{0.947}O} + \left\{ n^A_C + (C/Fe)^m \right\} \cdot H°_{1200 \atop C} + n^B_O \cdot \left(\tfrac{1}{2} \cdot H°_{T \atop O^B_2} + \frac{0.79}{0.21} \cdot \tfrac{1}{2} \cdot H°_{T \atop N^B_2} \right)$$

$$= H°_{1800 \atop Fe_l} + (C/Fe)^m \cdot H_{1800 \atop [C]}$$

$$+ n^A_C \cdot \left[\left\{ 2 - (O/C)^{gwrz} \right\} \cdot H°_{1200 \atop CO} + \left\{ (O/C)^{gwrz} - 1 \right\} \cdot H°_{1200 \atop CO_2} \right]$$

$$+ n^B_O \cdot \frac{0.79}{0.21} \cdot \tfrac{1}{2} \cdot H°_{1200 \atop N_2}. \tag{8.1}$$

8.2 The Demand-Supply Form of the Enthalpy Equation

The concepts of heat demand $\mathscr{D}$ and heat supply S were introduced in Chapter 5. As is pointed out there, a demand–supply form of the enthalpy equation has the advantage that increases in heat demand can simply be added to $\mathscr{D}$, and increases in enthalpy supply (e.g. due to an increased blast temperature) can simply be added to $\bar{S}$.

Equation (8.1) is altered into this form by making the substitutions

$$1.06 \cdot H°_{1200 \atop Fe_{0.947}O} = 1.06 \cdot H^f_{1200 \atop Fe_{0.947}O} + H°_{1200 \atop Fe_s} + 1.06 \cdot \tfrac{1}{2} \cdot H°_{1200 \atop O_2},$$

$$H°_{1200 \atop CO} = H^f_{1200 \atop CO} + H°_{1200 \atop C} + \tfrac{1}{2} \cdot H°_{1200 \atop O_2},$$

$$H°_{1200 \atop CO_2} = H^f_{1200 \atop CO_2} + H°_{1200 \atop C} + H°_{1200 \atop O_2}$$

where the H^f_{1200} terms are the enthalpies of formation of the compounds at 1200 K from elements at 1200 K. H^f_{1200} values, obtained from the JANAF tables (Stull, 1970), are listed in Appendix V.

Substituting these equations into equation (8.1) and changing the signs of all terms:

$$-1.06 \cdot H^f_{1200 \atop Fe_{0.947}O} + [H°_{1800 \atop Fe_l} - H°_{1200 \atop Fe_s}] + (C/Fe)^m \cdot [H_{1800 \atop [C]} - H°_{1200 \atop C}]$$

Enthalpy Balance

$$= -n_C^A \cdot \left[\left\{ 2 - (O/C)^{gwrz} \right\} \cdot H^f_{1200 \atop CO} + \left\{ (O/C)^{gwrz} - 1 \right\} \cdot H^f_{1200 \atop CO_2} \right]$$

$$+ n_O^B \cdot \left\{ (\tfrac{1}{2} \cdot H^\circ_{T_{O_2^B}} - \tfrac{1}{2} \cdot H^\circ_{1200 \atop O_2}) + \frac{0.79}{0.21} \cdot (\tfrac{1}{2} \cdot H^\circ_{T_{N_2^B}} - \tfrac{1}{2} \cdot H^\circ_{1200 \atop N_2}) \right\}$$

$$- \tfrac{1}{2} \cdot H^\circ_{1200 \atop O_2} \cdot \left\{ n_C^A \cdot (O/C)^{gwrz} - n_O^B - 1.06 \right\}. \tag{8.2}$$

This equation looks complicated but it will be seen later that the last term disappears when the enthalpy and stoichiometric equations are combined. The left-hand side of the equation is the heat demand and the first two terms of the right-hand side are the heat supply. Simplifying, equation (8.2) becomes

$$D^{wrz} = S^{wrz}$$
(wustite reduction zone heat supply)

$$= -n_C^A \cdot \left[\left\{ 2 - (O/C)^{gwrz} \right\} \cdot H^f_{1200 \atop CO} + \left\{ (O/C)^{gwrz} - 1 \right\} \cdot H^f_{1200 \atop CO_2} \right]$$

$$+ E^B \cdot n_O^B - \tfrac{1}{2} \cdot H^\circ_{1200 \atop O_2} \cdot \left\{ n_C^A \cdot (O/C)^{gwrz} - n_O^B - 1.06 \right\} \tag{8.3}$$

where:

$$E^B = (\tfrac{1}{2} \cdot H^\circ_{T_{O_2^B}} - \tfrac{1}{2} \cdot H^\circ_{1200 \atop O_2}) + \frac{0.79}{0.21} \cdot (\tfrac{1}{2} \cdot H^\circ_{T_{N_2^B}} - \tfrac{1}{2} \cdot H^\circ_{1200 \atop N_2})$$

which is the enthalpy of the dry blast air in excess of that which it would have at 1200 K.

8.3 Numerical Development

Numerical substitutions which can be inserted into equation (8.3) are

$$H^f_{1200 \atop CO} = -113\,000 \text{ kJ (kg mole of CO)}^{-1},$$

$$H^f_{1200 \atop CO_2} = -395\,000 \text{ kJ (kg mole of CO}_2)^{-1},$$

$$(O/C)^{gwrz} = 1.3 \qquad \text{(Section 7.3),}$$

from which

$$D^{wrz} = S^{wrz}$$

$$= n^A_C \cdot \left[\{2 - 1.3\} \cdot 113\,000 + \{1.3 - 1\} \cdot 395\,000 \right]$$

$$+ E^B \cdot n^B_O - \tfrac{1}{2} \cdot H^\circ_{1200 \atop O_2} \cdot \left\{ n^A_C \cdot 1.3 - n^B_O - 1.06 \right\} \qquad (8.4)$$

$$= n^A_C \cdot \left[198\,000 \right] + E^B \cdot n^B_O$$

$$- \tfrac{1}{2} \cdot H^\circ_{1200 \atop O_2} \cdot \left\{ 1.3 \cdot n^A_C - n^B_O - 1.06 \right\}. \qquad (8.4a)$$

8.3.1 The Value of D^{wrz}

D^{wrz} will vary, of course, with the amount of gangue and slag in the furnace, the rate of radiative and convective heat loss from the bottom segment and the enthalpy demands of subsidiary reactions in the bottom segment. However, for the simplest case in which the only demand is for producing a 5% carbon pig iron at 1800 K, the wustite reduction zone heat demand is

$$D^{wrz} = -1.06 \cdot H^f_{1200 \atop \underline{Fe_{0.947}O}} + [H^\circ_{1800 \atop Fe_l} - H^\circ_{1200 \atop Fe_s}]$$

$$+ (C/Fe)^m \cdot [H_{1800 \atop [C]} - H^\circ_{1200 \atop C}]. \qquad \text{(from equation (8.2))}$$

The numerical values of the terms in this equation are

H^f_{1200} $Fe_{0.947}O$ = $-265\,000$ kJ (kg mole)$^{-1}$,

H°_{1800} Fe_l $-$ H°_{1200} Fe_s = $38\,500$ kJ (kg mole)$^{-1}$,

H_{1800} $[C]$ $-$ H°_{1200} C = $44\,000$ kJ (kg mole)$^{-1}$ (including the heat of mixing carbon and iron, Chapter 12),

$(C/Fe)^m$ = 0.25,

from which the heat demand D^{wrz} is $330\,000$ kJ (kg mole of product Fe)$^{-1}$. Other demands, e.g. for slag heating and melting, are simply added to this base value (Problem 8.2). Wustite reduction zone heat demands which more truly represent industrial blast-furnace conditions are calculated in Chapter 12.

8.3.2 Values of E^B

E^B represents the enthalpy which the dry blast air has in excess of that which it would have at the thermal reserve temperature. It is positive when $T_B > 1200$ K; zero when $T_B = 1200$ K; and negative when $T_B < 1200$ K. Numerical values of E^B are tabulated in Appendix VII.

8.4 Summary

The concept of dividing the blast furnace into two segments has been continued in this chapter and a second steady-state enthalpy equation in terms of the variables n^A_C and n^B_O has been developed for the bottom segment. Continuity between segments was maintained while developing equation (8.4) and for this reason the operating parameters calculated with it are applicable to the whole furnace as well as to the bottom segment.

As will be shown in the next chapter, this enthalpy equation and bottom segment stoichiometric equation (7) can be combined to define the operating requirements (n^A_C and n^B_O) of any blast furnace.

Reference

Stull, D. R., Prophet, H. *et al.* (1970) *JANAF Thermochemical Tables* 2nd edition, United States Department of Commerce, Document NSRDS-NBS 37, Washington, June 1971.

Chapter 8 Problems. *Wustite Reduction Zone Enthalpy Equation; Heat Demand; Blast Enthalpy*

8.1 Prove that equation (8.2) follows from equation (8.1). Do not ignore any terms.

8.2 Calculate the wustite reduction zone heat demand for the conditions cited in Problem 5.1 (a,b,c,d,e). Include the heat of mixing of carbon in iron (Section 8.3) but neglect heats of mixing Mn and Si in iron.
Assume:
  (i) a thermal reserve temperature of 1200 K;
  (ii) that the compounds reduced in the wustite reduction zone are $Fe_{0.947}O$, MnO and SiO_2;
  (iii) that four-fifths of the radiative and convective heat losses take place in the bottom segment.

What is represented by the differences between the answers to Problem 5.1 and the answers to this problem?

8.3 Blast enters the Burns Harbour furnace at 1350 K. Calculate the value of E^B for this blast and state precisely what this term represents.

8.4 The temperature of blast in old blast-furnace installations is often below 1200 K due to a lack of adequate stove capacity. To show the effect of this, calculate the value of E^B for 1050 K blast. What is the physical significance of your answer?

8.5 The enthalpy equation for the wustite reduction zone is complicated by the term:

$$-\tfrac{1}{2} H^\circ_{1200_{O_2}} \cdot \left\{ 1.3 \cdot n_C^A - n_O^B - 1.06 \right\}.$$

Calculate, using Problem 7.1 as an example, the value of this term. Can you make a general statement about this term based upon equation (7)?

CHAPTER 9

Combining Bottom Segment Stoichiometry and Enthalpy Equations: a priori *Calculation of Operating Parameters*

Our efforts towards developing a powerful mathematical description of the blast furnace have led to

(a) a stoichiometric equation for the whole furnace (equation (4));
(b) a stoichiometric equation for the bottom segment of a conceptually divided furnace:

$$n_O^B + 1.06 = 1.3 \cdot n_C^A; \qquad (7)$$

(c) an enthalpy equation for the bottom segment of the conceptually divided furnace:*

$$D^{wrz} = S^{wrz}$$
(enthalpy supply in bottom segment)
$$= n_C^A \cdot (198\,000) + E^B \cdot n_O^B - \tfrac{1}{2} \cdot H^\circ_{1200,\,O_2} \cdot (1.3 \cdot n_C^A - n_O^B - 1.06). \qquad (8.4a)*$$

As a first step in this chapter, equation (8.4a) can be simplified immediately because its last term is always zero, as shown by equation (7).

*Where it will be remembered

D^{wrz} = heat demand for reduction, heating and melting in the bottom segment
1.06 = $(O/Fe)^{xwrz}$ (wustite),
1.3 = $(O/C)^{gwrz}$ ($CO/CO_2/Fe/Fe_{0.947}O$ equilibrium at 1200 K),
198 000 = $-\left[\left\{2-(O/C)^{gwrz}\right\} \cdot H^f_{1200,\,CO} + \left\{(O/C)^{gwrz} - 1\right\} \cdot H^f_{1200,\,CO_2}\right].$

92   The Iron Blast Furnace

Elimination of this term gives a final bottom segment enthalpy equation:
$$D^{wrz} = S^{wrz} = n_C^A \cdot (198\,000) + E^B \cdot n_O^B. \tag{8}$$

Equations (7) and (8) assume (i) the existence of chemical and thermal reserve zones in the blast furnace; (ii) a thermal reserve temperature of 1200 K; (iii) a chemical reserve carbonaceous gas composition of 70% CO, 30% CO_2 (i.e. the gas composition which would be at equilibrium with $Fe_{0.947}O$ and Fe at 1200 K). These assumptions are re-examined in Chapter 10.

9.0.1 Power of the Equations

Equations (7) and (8) indicate that once the enthalpy of the blast, E^B, and the heat demand of the wustite reduction zone, D^{wrz}, are specified we have two equations with two unknowns (n_O^B and n_C^A). Thus the operating requirements of the process, n_O^B (oxygen in blast air per mole of product Fe) and n_C^A (carbon taking part in reduction and heating per mole of product Fe), are uniquely defined.

For calculation purposes, equations (7) and (8) may be combined by substituting
$$n_O^B = 1.3 \cdot n_C^A - 1.06 \tag{7}$$

into equation (8) which becomes
$$D^{wrz} = S^{wrz} = n_C^A \cdot (198\,000) + E^B \cdot (1.3 \cdot n_C^A - 1.06)$$
$$= n_C^A \cdot (198\,000 + 1.3 \cdot E^B) - 1.06 \cdot E^B. \tag{9}$$

Equation (9) provides a straightforward method for calculating n_C^A from numerical values of D^{wrz} and E^B. Blast oxygen, n_O^B, is then readily determined from n_C^A by means of equation (7). Equations (7) and (8) can, of course, be solved by any simultaneous equation method but equation (9) has the advantage* that it illustrates clearly how n_C^A is affected by changes in D^{wrz} and E^B.

*Likewise the equation
$$1.3 \cdot D^{wrz} = n_O^B \cdot (198\,000 + 1.3 \cdot E^B) + (1 \cdot 06) \cdot (198\,000)$$
specifies the effects of D^{wrz} and E^B on n_O^B.

Bottom Segment Stoichiometry and Enthalpy Equations

TABLE 9.1

Item	Specification	Quantity		Model variable
		kg per tonne of Fe	kg moles per tonne of Fe	
Fe		1000	17.9	
Iron oxide entering wustite reduction zone	$Fe_{0.947}O$			$(O/Fe)^{xwrz} = 1.06$
Pig iron	5% C			$(C/Fe)^m = 0.25$
Blast temperature				$T_B = 1400$ K
Oxygen from blast air				$n_O^B = ?$
Active carbon				$n_C^A = ?$
Total carbon				$n_C^t = (C/Fe)^m + n_C^A$
Heat demand of wustite reduction zone				$D^{wrz} = 330\,000$ kJ per kg mole of product Fe

9.1 Example Calculations

The uses and some of the implications of equations (9) and (7) are shown by the following example in which the carbon and blast required to produce 1 tonne of Fe under minimum heat-demand conditions are calculated. The conditions for this illustrative calculation are shown in Table 9.1. The wustite reduction zone heat demand in this example is the minimum for iron oxide reduction. It is (Section 8.3) the enthalpy required to produce a liquid 5% C pig iron (1800 K) from $Fe_{0.947}O$ (1200 K), i.e. there are no gangue or slag, no minor reactions and no radiative or convective heat losses.

To start the calculation it will be remembered that E^B represents the enthalpy in the blast above that which it would have at 1200 K. Numerically

$$E^B = \tfrac{1}{2} \cdot \left\{ [H^{\circ}_{1400} - H^{\circ}_{1200}]_{O_2} + 3.76 \cdot [H^{\circ}_{1400} - H^{\circ}_{1200}]_{N_2} \right\}$$

$$= \tfrac{1}{2} \cdot \{[37\,000 - 30\,000] + 3.76 \cdot [35\,000 - 28\,000]\}$$

$$= 17\,000 \text{ kJ (kg mole of O)}^{-1}. \qquad \text{(data, Appendix VII)}$$

Insertion of this value of E^B and the specified value of 330 000 kJ for D^{wrz} into equation (9) gives

$$330\,000 = n^A_C \cdot \{198\,000 + 1.3 \cdot (17\,000)\} - 1.06 \cdot (17\,000)$$

from which

$$n^A_C = 1.58 \text{ moles of C per mole of product Fe.}$$

Similarly, equation (7) becomes

$$n^B_O + 1.06 = 1.3 \cdot (1.58)$$

and

$$n^B_O = 0.99 \text{ moles of O per mole of product Fe.}$$

Thus the input requirements for this operation are as shown in Table 9.2 and the top gas composition (hematite charge) is, from equation (4) and Section 4.3; 19 vol% CO, 27% CO_2 and 54% N_2.

TABLE 9.2

Item	Specification	Quantity		Model variable (kg moles per kg mole of product Fe)
		kg per tonne of Fe	kg moles per tonne of Fe	
Oxygen from blast air	285 (945 Nm³ of air blast)		17.7 (O)	$n_O^B = 0.99$
Active carbon	340		28.3	$n_C^A = 1.58$
Total carbon	393		32.8	$n_C^i = (C/Fe)^m + n_C^A$ $= 0.25 + 1.58 = 1.83$

9.2 Implications of the Equations

9.2.1 Effects of an Increased Heat Demand

Equations (9) and (7) show clearly that an increase in wustite reduction zone heat demand (for minor endothermic reactions, for heating and melting of gangue and slag and for convective and radiative heat losses) inevitably leads to increased blast-furnace demands for carbon (n_C^A in equation (9)) and blast oxygen (n_O^B in equation (7)). Thus large slag falls (per tonne of Fe) and heat losses lead to high coke demands and low furnace productivities.

9.2.2 Effects of Blast Temperature

Increases in blast temperature (i.e. increases in E^B) result in smaller requirements for carbon and blast oxygen. This can be seen by calculating n_C^A and n_B^O for increasing values of E^B (equations (9) and (7)) or graphically as is described in Section 9.5.

9.2.3 Ignoring the Nature of the Charge

It is readily noticeable that equations (9) and (7) ignore (i) the nature of the oxide in the charge (i.e. Fe_2O_3 or Fe_3O_4) and (ii) the heat demands of the top segment of the furnace.

This is consistent with the views that:

(a) stoichiometrically and thermodynamically there is always enough CO remaining after wustite reduction to produce the necessary quantity of wustite from Fe_2O_3 and Fe_3O_4;
(b) the gas rising from the bottom segment into the top carries sufficient enthalpy to heat the charge, to offset endothermic reduction reactions, and to account for convective and radiative heat losses from the top segment of the furnace.

These views are re-examined in Chapter 10.

Bottom Segment Stoichiometry and Enthalpy Equations

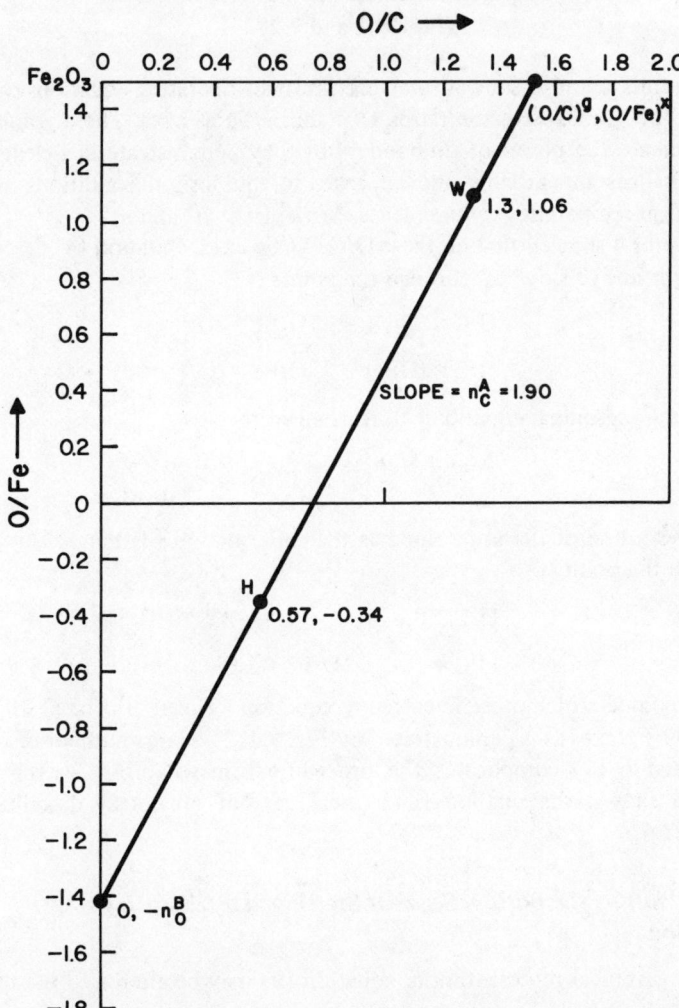

Fig. 9.1. Blast-furnace *operating line* as fixed by enthalpy point H and wustite reduction point W (Section 9.3). The numerical values are for the blast-furnace operation described in Section 9.4.

The Iron Blast Furnace

9.3 Graphical Representation of the Equations
(Figs. 9.1 and 9.2)

Chapters 4 and 6 showed that blast-furnace-operating equations can be plotted on graphs with the ratios O/C and O/Fe as axes. These graphs do not increase the power of the model but they demonstrate in a clear way how changes in enthalpy and approach-to-equilibrium conditions affect blast-furnace-operating requirements, particularly n_C^A and n_O^B.

Chapter 4 showed that on these O/C : O/Fe axes, equation (4) describes a straight line of slope n_C^A through the points

$$O/C = 0, \qquad O/Fe = -n_O^B,$$
$$O/C = (O/C)^g, \qquad O/Fe = (O/Fe)^x.$$

By similar reasoning, equation (7), rearranged to

$$1.06 - (-n_O^B) = n_C^A \cdot (1.3 - \underset{zero}{0})$$

describes a line of the same slope as that of equation (4) but in this case through the points

$$O/C = 0, \qquad O/Fe = -n_O^B,$$
$$O/C = 1.3, \qquad O/Fe = 1.06. \qquad \text{(point W, Fig. 9.1)}$$

Combined stoichiometric/enthalpy equation (9) can also be plotted on O/C : O/Fe axes as is demonstrated by Fig. 9.1. The large number of terms in equation (9) complicates the procedure somewhat but, as the next section shows, the method is identical to that previously described in Chapter 6.

9.3.1 Plotting the Bottom Segment Stoichiometry/Enthalpy Equation

For graphical representation, equation (9) may be altered to the form*

$$D^{wrz} = n_C^A \cdot [1.3 \cdot \{282\,000 + E^B\} - 169\,000] - 1.06 \cdot E^B \qquad (9a)$$

*$282\,000 = H^f_{1200} - H^f_{1200}; \quad 169\,000 = 2H^f_{1200} - H^f_{1200}.$
$\qquad\qquad\quad\;\text{CO}\quad\;\;\text{CO}_2 \qquad\qquad\qquad\;\;\text{CO}\quad\;\;\text{CO}_2$

Bottom Segment Stoichiometry and Enthalpy Equations

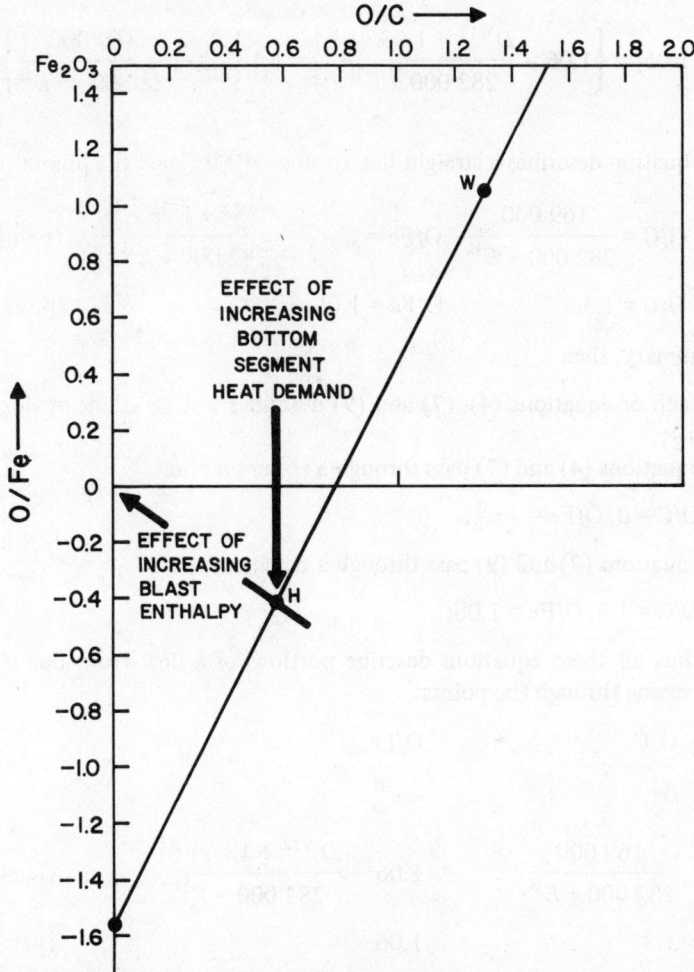

Fig. 9.2. Effects of blast enthalpy and wustite reduction zone heat demand on the slope and intercepts of the blast-furnace *operating line*. Increases in blast enthalpy are shown to flatten the slope and raise the left intercept, which are equivalent (Section 9.5) to decreasing the carbon and blast requirements of the process (per tonne of Fe). Increases in heat demand cause the opposite.

which for plotting purposes may be further rearranged to

$$1.06 - \left\{1.06 - \frac{D^{\mathrm{wtz}} + 1.06 \cdot E^{\mathrm{B}}}{282\,000 + E^{\mathrm{B}}}\right\} = n_{\mathrm{C}}^{\mathrm{A}}\left\{1.3 - \frac{169\,000}{282\,000 + E^{\mathrm{B}}}\right\}. \quad (9b)$$

This equation describes a straight line of slope $n_{\mathrm{C}}^{\mathrm{A}}$ through the points

$$O/C = \frac{169\,000}{282\,000 + E^{\mathrm{B}}}, \quad O/Fe = 1.06 - \frac{D^{\mathrm{wtz}} + 1.06\,E^{\mathrm{B}}}{282\,000 + E^{\mathrm{B}}} \quad \text{(point H)},$$

$$O/C = 1.3, \qquad O/Fe = 1.06 \qquad \text{(point W)}.$$

In summary, then:

(a) each of equations (4), (7) and (9) describes a straight line of slope $n_{\mathrm{C}}^{\mathrm{A}}$;

(b) equations (4) and (7) pass through a common point:

$O/C = 0$, $O/Fe = -n_{\mathrm{O}}^{\mathrm{B}}$;

(c) equations (7) and (9) pass through a common point:

$O/C = 1.3$, $O/Fe = 1.06$;

(d) thus all three equations describe portions of a line with slope $n_{\mathrm{C}}^{\mathrm{A}}$ passing through the points:

O/C	O/Fe	
0	$-n_{\mathrm{O}}^{\mathrm{B}}$,	
$\dfrac{169\,000}{282\,000 + E^{\mathrm{B}}}$	$1.06 - \dfrac{D^{\mathrm{wtz}} + 1.06 \cdot E^{\mathrm{B}}}{282\,000 + E^{\mathrm{B}}}$	(point H),
1.3	1.06	(point W),
$(O/C)^{\mathrm{g}}$	$(O/Fe)^{x}$.	

The positions of these points are slightly altered by hydrocarbon and/or oxygen injection into the furnace. However, the effects of these additions are readily included on the diagram as is shown in detail in Chapter 11.

9.4 A Graphical Calculation

As a graphical example, consider that the bottom segment heat demand (D^{wrz}) of the operation described in Section 9.1 is increased by 70 000 kJ to a more realistic value (including flux, slag, heat losses, minor reactions) of 400 000 kJ per kg mole of product Fe. The iron oxide charge remains the same as do the temperatures and compositions of the blast and the product iron.

The data for this altered operation are presented in Table 9.3.

The object of the calculation is to determine n_C^A, n_C^i, n_O^B and the top-gas composition, $(O/C)^g$. The first point to be plotted (Fig. 9.1) is point W:

$$O/C = 1.3, \qquad O/Fe = 1.06.$$

The second point is point H:

$$O/C = \frac{169\,000}{282\,000 + E^B}, \qquad O/Fe = 1.06 - \frac{D^{wrz} + 1.06 \cdot E^B}{282\,000 + E^B}$$

which from Table 9.3 is

$$O/C = 0.57, \qquad O/Fe = -0.34.$$

A line is drawn between these points, and its slope is determined to be

$$n_C^A = 1.90.$$

The blast oxygen requirement is then calculated by extending the operating line to the point

$$O/C = 0, \qquad O/Fe = -n_O^B = ?$$

from which

$$n_O^B = 1.41.$$

Similarly, the composition of the top gas is obtained by extending the operating line to the point

$$O/C = (O/C)^g = ?, \qquad O/Fe = (O/Fe)^x = 3/2,$$

TABLE 9.3

Item	Specification	Quantity		Model variable (kg moles per kg mole of product Fe)	
		kg per tonne of Fe	kg moles per tonne of Fe		
Fe		1000	17.9		
Input iron oxide	Fe_2O_3			$(O/Fe)^x$	$= 3/2$
Iron oxide entering wustite reduction zone	$Fe_{0.947}O$			$(O/Fe)^{x,wrz}$	$= 1.06$
Pig iron	5% C			$(C/Fe)^m$	$= 0.25$
Blast temperature				T_B	$= 1400$ K
Blast enthalpy	1400 K			E^B	$= 17\,000$ kJ per kg mole of O (Section 9.1)
Heat demand of wustite reduction zone				D^{wrz}	$= 400\,000$ kJ per kg mole of product Fe

TABLE 9.4

Item	Specification	Quantity		Model variable (kg moles per kg mole of product Fe)
		kg per tonne of Fe	kg moles per tonne of Fe	
Oxygen from blast air		404 (1350 Nm³ of air)	25.2 (O)	$n_O^B = 1.41$
Active carbon				$n_C^A = 1.90$
Total carbon		462	38.5	$n_C^i = (C/Fe)^m + n_C^A$ $= 0.25 + 1.90$ $= 2.15$

from which

$$(O/C)^g = 1.53,$$

i.e. $X^g_{CO_2} = 0.53,$ $X^g_{CO} = 0.47.$

The complete gas analysis, including N_2, can be calculated as in Section 4.3.

The completed table of operating requirements for this example is as shown in Table 9.4.

9.5 Characteristics of the Operating Line

The operating equations can be plotted on O/C : O/Fe axes and the (4), (7) and (9) are (Fig. 9.2):

(a) variations in operating conditions (flux and slag load, heat losses, enthalpy in blast) cause the line to rotate about point W (O/C = 1.3, O/Fe = 1.06) which, assuming that equilibrium is closely approached in the chemical reserve, has a fixed position independent of the operating parameters of the furnace;

(b) the position of the line as it rotates about point W is determined by the enthalpy demand of the bottom segment and the enthalpy content of the blast, both of which establish the coordinates of point H.

9.5.1 Effects of Bottom Segment Heat Demand on Carbon and Oxygen Requirements

Section 9.3.1 has shown that the coordinates of point H are:

$$(O/C)_H = \frac{169\,000}{282\,000 + E^B}, \quad (O/Fe)_H = 1.06 - \frac{D^{WIZ} + 1.06 \cdot E^B}{282\,000 + E^B}$$

$$= \frac{1.06 \cdot 282\,000 - D^{WIZ}}{282\,000 + E^B},$$

(9.1a, 9.1b)

Bottom Segment Stoichiometry and Enthalpy Equations

from which it is readily seen that the vertical position of point H is directly affected by the bottom segment heat demand D^{wrz} but that its horizontal position is not.

It can be seen further that an increase in D^{wrz} causes point H to move downward vertically (Fig. 9.2) which in turn:

(a) causes the operating line to rotate counter-clockwise around fixed point W;
(b) leads to a steeper operating line (i.e. a larger active carbon [n_C^A] requirement) and a deeper O/Fe intercept (i.e. a larger blast oxygen [n_O^B] requirement).

Of course a decreased D^{wrz} results in opposite effects.

9.5.2 Effects of Blast Enthalpy

The coordinates of point H given immediately above can be represented by the equation:

$$\frac{(O/Fe)_H}{(O/C)_H} = \frac{1.06 \cdot 282\,000 - D^{wrz}}{169\,000}$$

or $\quad (O/Fe)_H = \dfrac{1.06 \cdot 282\,000 - D^{wrz}}{169\,000} \cdot (O/C)_H \quad$ (9.2)

which can be seen to describe (for any prescribed value of D^{wrz}) a straight line through the O/C = 0 : O/Fe = 0 origin.

Furthermore, increases in E^B cause both $(O/C)^H$ and $(O/Fe)^H$ to approach zero (equations (9.1a,b)), thereby moving point H towards the origin, with the net results (Fig. 9.2) that:

(a) the operating line rotates clockwise around point W and becomes less steep;
(b) n_C^A and n_O^B both decrease.

Increases in blast enthalpy (i.e. increases in blast temperature) result, therefore, in decreased requirements for both carbon and blast.

It will be remembered that E^B represents the enthalpy content of the blast above that which it would have at 1200 K. E^B can, therefore, have negative values. By similar reasoning to that above, the colder the blast (the more negative the value of E^B) the larger will be the carbon and oxygen requirements of the operation.

9.6 Summary

This chapter has completed our model in its most basic form. A final wustite reduction zone enthalpy equation has been developed and this plus stoichiometric equations for (i) the whole furnace and (ii) the wustite reduction zone, have been shown to completely define how a blast furnace must operate. Two operating parameters must be known before the definition is complete: blast enthalpy and wustite reduction zone heat demand D^{wrz}. Once these are specified (from blast temperature, minor reactions, quantities of gangue and slag, etc.), the major operating requirements of the furnace are readily calculated, both analytically and graphically.

The operating equations can be plotted on O/C : O/Fe axes and the resulting graphs can be used to demonstrate clearly the effects of altered heat demands and blast temperatures on furnace-operating parameters. High heat demands are shown to require large carbon and oxygen inputs while high blast temperatures have the opposite effect.

The remainder of the text is devoted to widening the scope of the model by incorporating additional blast-furnace variables into the operating equations. The premises, predictions and potential applications of the model are also examined and evaluated.

Chapter 9 Problems. A Priori *Prediction of Blast-furnace-operating Parameters*

9.1 A blast furnace is operating with dry, hot air blast at 1450 K. The charge consists of hematite (5% SiO_2) sinter, CaO and coke (assume pure C). Its product metal contains 5 wt.% carbon (ignore other impurities) and its slag may be considered to consist only of CaO and SiO_2 (CaO/SiO_2 wt. ratio = 1.2). Wustite reduction zone convective and radiative heat losses are similar to those calculated in Problem 8.2.

Calculate analytically for this operation:

(a) the wustite reduction zone heat demand;
(b) E^B;

Bottom Segment Stoichiometry and Enthalpy Equations

    (c)   the carbon requirement (per tonne of Fe);
    (d)   the blast requirement (per tonne of Fe);
    (e)   the top gas composition.

9.2 A shortage of high-grade material forces the operators of the furnace described in Problem 9.1 to use a low-grade (15% SiO_2) stockpiled sinter. To make matters worse; (i) the sulphur content of the coke increases, necessitating an increase in the CaO/SiO_2 weight ratio to 1.3; and (ii) a stove goes off line causing a decrease of blast temperature to 1250 K.

Calculate graphically:

    (a)   the carbon requirement,
    (b)   the blast requirement,

for this worsened operation. Assume that the heat demand of the slag (per kg) is the same as in Problems 8.2 and 9.1.

9.3 Prepare a computer programme which will calculate:

    (a)   the blast requirement,
    (b)   the carbon requirement,
    (c)   the top gas composition,

for any blast-furnace operation from given values of wustite reduction zone heat demand (D^{wrz}), blast temperature and % C in hot metal. Use, as far as possible, your programmes from Problems 4.8 and 7.4. Employ the new programme to check your answers to Problems 9.1 and 9.2.

9.4 Instruct the Problem 9.3 programme to plot the operating line for any specified blast-furnace operation, given D^{wrz} and blast temperature. The coordinates of points H and W; the intercepts at O/C = 0 and O/Fe = (O/Fe)^x; and the slope of the operating line should be labelled.

9.5 Silicon and manganese are reduced from SiO_2 and MnO below the chemical reserve. Show where the quantities of Si and Mn in the product metal (as represented by the molar ratios $(Si/Fe)^m$ and $(Mn/Fe)^m$) would appear on the stoichiometric diagram.

CHAPTER 10

Testing of the Mathematical Model and a Discussion of its Premises

The development of equation (7)

$$n_O^B + 1.06 = 1.3 \cdot n_C^A$$

and equation (8):*

$$D^{wrz} = S^{wrz}$$
$$= n_C^A \cdot (198\,000) + E^B \cdot n_O^B$$

has completed the basic mathematical description of the blast furnace. As was shown in Chapter 9, these two equations provide the basic tools for calculating the coke and blast requirements for any given furnace charge. They are useful in their basic form and they can be expanded further (Chapters 11, 12) to include all other operating parameters of the process, e.g. tuyère injectants, moisture, slag components, heat losses, etc.

However, before the model is embellished, it requires testing and discussion, which is the purpose of this chapter. Specifically, the model is tested for (i) thermal validity, (ii) stoichiometric validity and (iii) thermodynamic (equilibrium) validity. Accuracies of the model predictions are also discussed.

10.1 Testing for Thermal Validity

The basic test, which the model must meet for it to be thermally valid, is that the temperature which it predicts for the gas leaving the top of the

*Where $198\,000 = 0.7 \cdot (-H^f_{1200})_{CO} + 0.3 \cdot (-H^f_{1200})_{CO_2}$.

Testing of the Mathematical Model

furnace must not be lower than the temperature which it has assigned to the incoming solid charge. Only in this way can the model represent the fact that the ascending gas can never become colder than the solids which it heats.

Expressed in arithmetical terms, the basic test for thermal validity is

$$T_g \geqslant 298 \text{ K}. \tag{10.1}$$

As will be shown by the following top-gas temperature calculation, this thermal restriction is always satisfied by equations (7) and (8) in spite of the fact that these equations were developed for the bottom segment of the furnace only. This is because the gas rising from the thermal reserve zone always contains sufficient enthalpy to heat the charge and to provide enthalpy for higher oxide reduction.

10.1.1 Determining Top-gas Temperature: Complete Enthalpy Balance for the Furnace

A simplified whole-furnace enthalpy balance is described in Chapter 5. To represent actual operating conditions it must now be expanded to include (i) blast enthalpy; (ii) enthalpy in top gas; and (iii) enthalpy of carbon-in-iron. These factors are most clearly included into the enthalpy balance by incorporating them in equation (5.3):

$$\mathcal{D} = S = n_C^A [\{2 - (O/C)^g\} \cdot 111\,000 + \{(O/C)^g - 1\} \cdot 394\,000] \tag{5.3}$$

which becomes:

$$\mathcal{D} = n_C^A [\{2 - (O/C)^g\} \cdot 111\,000 + \{(O/C)^g - 1\} \cdot 394\,000]$$

$$+ n_O^B \left[\tfrac{1}{2} [H_{T_B}^\circ]_{O_2} + \frac{0.79}{0.21} \cdot \tfrac{1}{2} [H_{T_B}^\circ]_{N_2} \right]$$

$$- n_C^A \left[\{2 - (O/C)^g\} \cdot [H_{T_g}^\circ - H_{298}^\circ]_{CO} + \{(O/C)^g - 1\} \cdot [H_{T_g}^\circ - H_{298}^\circ]_{CO_2} \right]$$

$$- n_O^B \cdot \frac{0.79}{0.21} \cdot \tfrac{1}{2} [H_{T_g}^\circ]_{N_2}. \tag{10}$$

The Iron Blast Furnace

It can be seen that the first line of equation **(10)** is precisely equation **(5.3)**. The second line represents the enthalpy in blast at any blast temperature T_B. The third line represents the sensible heat (above 298 K) carried out of the furnace in the top gas.

The heat demand in equation **(10)** is that for the whole furnace, i.e.

$$\mathscr{D} = -\tfrac{1}{2} H^f_{298} + H^\circ_{1800} + (C/Fe)^m \cdot H_{1800} \qquad \text{kJ (kg mole of Fe)}^{-1}.$$
$$\quad\;\; Fe_2O_3 \qquad\;\; Fe_l \qquad\qquad\quad [C] \qquad\qquad\qquad\qquad\qquad (10.2)$$

It has been slightly altered from the equation (5.3) demand in that it now realistically includes the enthalpy of the carbon in the iron product.

10.2 Top-gas Temperature Calculation

Calculation of top-gas temperatures from calculated values of n^A_C and n^B_O can be illustrated by continuing with the example problem on page 94. The specifications for this furnace were as shown in Table 10.1 while, from equation (10.2), its whole-furnace heat demand* is

$$\mathscr{D} = -\tfrac{1}{2} \cdot (-826\,000) + 73\,000 + 0.25 \cdot (60\,000) = 501\,000$$
$$\text{kJ (kg mole of product Fe)}^{-1}.$$

Once $\mathscr{D}$ has been specified, all of the operational variables in equation **(10)** are in hand (i.e. $\mathscr{D}$, n^A_C, $(O/C)^g$, n^B_O, T_B) and the top-gas temperature T_g is readily calculated. The simplest method of calculation is to represent the terms

$$[H^\circ_{T_g} - H^\circ_{298}]_{CO},\; [H^\circ_{T_g} - H^\circ_{298}]_{CO_2} \text{ and } [H^\circ_{T_g}]_{N_2}$$

---

* $H_{1800} = H^\circ_{1800} + H^M_C$ \qquad\qquad (Section 12.2.4)
\quad [C] \qquad C \qquad\; 1800

$\qquad = 30\,000 + 30\,000$ kJ (kg mole of C)$^{-1}$.

$\mathscr{D}$ will vary slightly with different input and output temperatures: 550 000 to 600 000 kJ per kg mole of Fe is typical of large, modern industrial blast furnaces.

TABLE 10.1

Item	Specification	Quantity		Model variable (kg moles per kg mole of product Fe)
		kg per tonne of Fe	kg moles per tonne of Fe	
Fe		1000	17.9	
Iron oxide	Fe_2O_3 (entering the furnace at 298 K)	1430	9.0	$(O/Fe)^x = 3/2$
Pig iron	5% C			$(C/Fe)^m = 0.25$
Blast temperature				$T_B = 1400$ K
Oxygen from blast	air	285	17.7 (O)	$n_O^B = 0.99$
Active carbon		340	28.3	$n_C^A = 1.58$
Total carbon		390	32.7	$n_C^i = (C/Fe)^m + n_C^A$
				$= 1.83$
Top-gas composition	$X_{CO} = 0.42$ $X_{CO_2} = 0.58$			$(O/C)^g = 1.58$

in equation (10) as linear functions of T_g, i.e.

$$[H°_{T_g} - H°_{298}]_{CO} = 30.2 \cdot T_g - 9\,100,$$

$$[H°_{T_g} - H°_{298}]_{CO_2} = 45.6 \cdot T_g - 14\,100$$

$$[H°_{T_g}]_{N_2} = 30.0 \cdot T_g - 9\,000$$

kJ (kg mole of gas)$^{-1}$ over the range 298 K to 800 K (Appendix VI),

which permits T_g to be determined directly. In this example T_g is 450 K, which is in line with the top gas temperatures of most modern blast furnaces (Table 1.1).

10.2.1 Comments on the Thermal Validity of the Model

There appear to be no practical conditions under which the equation (7, 8) operating parameters will not satisfy the top gas temperature ($T_g \geqslant 298$ K) test.* This is because:

(a) The heat demand for higher oxide reduction is nearly met by the heat supply from an equivalent amount of CO reacting to form CO_2, i.e.

$$0.44CO + \tfrac{1}{2}Fe_2O_3 \rightarrow 1.06Fe_{0.947}O + 0.44CO_2$$
$$\Delta H°_{298} = +6000 \text{ kJ (kg mole of product Fe)}^{-1}. \tag{2.1}$$

(b) The mass of gas rising into the top segment actually exceeds (Fig. 1.3) the mass of the equivalent descending charge due to O_2, N_2 and hydrocarbon additions through the tuyères.

(c) An increased top segment heat demand (from more gangue, flux, etc.) is always accompanied by an equivalent increase in the bottom segment heat demand (due to the same gangue, flux, etc.). The latter leads to (i) more carbon and blast being added to the furnace per tonne of product Fe (equations (7) and (8)); to (ii) the

*An exception to this would be if the blast entering a blast furnace were to be excessively enriched with pure oxygen. Kitaev (1967) states, for example, that the $T_g \geqslant 298$ K test will not be satisfied if the blast contains more than 30% O_2. Of course, 30% O_2 in blast will also cause an excessive tuyère flame temperature in most cases. These interrelationships can all be examined by means of the computer programme described in Chapter 13.

ascent of more hot gas into the top segment; and to (iii) maintenance of thermal validity. Convective and radiative heat losses in the top and bottom segments have similar offsetting effects.

It can be seen, then, that the operating parameters calculated by means of wustite reduction zone equations (7), (8) will always satisfy the top-gas temperature requirement. This can always be checked by means of the whole-furnace enthalpy equation (10) or the top segment enthalpy equation (equation (10) minus equation (8)).

10.3 Testing for Stoichiometric Validity

In addition to being thermally valid, the operating parameters (n_C^A, n_O^B) calculated by means of bottom segment equations (7) and (8) must satisfy the stoichiometric top-gas inequality:

$$(O/C)^g \leqslant 2. \tag{10.3}$$

This inequality states that each mole of carbon leaving the furnace in the gas cannot carry more than 2 moles of O with it, i.e. CO_2 is as far as carbon oxidation can proceed.

That inequality (10.3) is always satisfied by the predictions of the model is most easily shown by examining the example problem on page 94. This problem showed that (i) the lowest possible wustite reduction zone heat demand is 330 000 kJ per kg mole of product Fe and that (ii) with this minimum heat demand the predicted top-gas oxygen/carbon ratio is

$$(O/C)^g = 1.58.$$

It can be seen that this predicted ratio is far below the value of 2 required by Test 10.3.

The effects of heat demands larger than the above-mentioned minimum (i.e. including gangue, flux, heat losses, etc.) may be determined by combining equations (4), (7) and (8) to give

$$(O/C)^g = 1.3 + \frac{(O/Fe)^x - 1.06}{(D^{wrz}/198\,000)} \tag{10.4}$$

T^B assumed to be zero, i.e. $T_B = 1200$ K; $(O/Fe)^x = 3/2$ or $4/3$).

This equation shows clearly that increases in D^{wrz} above its minimum can only lead to values of $(O/C)^g$ which are less than the 1.58 of the above problem.

Thus the top-gas O/C ratio predicted by the operating equations will always be well below 2, and the test for stoichiometric validity will always be satisfied.

10.4 Testing for Thermodynamic Validity

Gas leaves the top of a blast furnace at a temperature between 400 and 500 K. At these temperatures, Fe_2O_3 or Fe_3O_4 can theoretically be reduced to Fe by carbonaceous gas of the approximate composition:

$$X_{CO} = 0.35, X_{CO_2} = 0.65 \quad \text{(thermodynamic data; Stull, 1970)},$$

i.e. by gas with an oxygen to carbon ratio of

$$(O/C)^g < 1.65.$$

Gases leaving a blast furnace at O/C ratios of less than 1.65 are, therefore thermodynamically capable of producing more iron than they actually do. The $(O/C)^g$ values resulting from the equation (7, 8) operating parameter are always less than 1.65 (usually between 1.4 and 1.6, Section 10.3) so that they always fall into this category.

This means that, taking the blast furnace as a whole, the equation (7, 8) parameters are satisfactory in the thermodynamic sense that they always predict an amount of iron which is within the thermodynamic capability of the top gas.

10.5 Validity of the Model Assumptions and Predictions

The previous three sections have shown that the equation (7, 8) predictions satisfy all the temperature, stoichiometric and thermodynamic requirements of the blast-furnace process. However, the question remains as to whether the assumptions of the mathematical equations give accurate

Testing of the Mathematical Model

rather than merely satisfactory predictions.

The equations are based, it will be remembered, on the assumptions that
  (a) the only form of iron descending through the chemical reserve into the wustite reduction zone is $Fe_{0.947}O$, i.e. that magnetite reduction to wustite is completed in the top segment but that no metallic iron is produced in the top segment;
  (b) coke is not gasified by the reaction:

$$CO_2 + C \longrightarrow 2CO \qquad (1.2)$$

  in or above the chemical reserve.

These assumptions are re-examined in the four following subsections.

10.5.1 Iron Descent into the Wustite Reduction Zone

Blast-furnace gas is capable (Section 10.4) of reducing Fe_2O_3 to Fe even as the gas leaves the furnace. This means that there is always a possibility of iron descending through the chemical reserve and into the wustite reduction zone. Such a descent of iron would lower the CO and heat demands of the wustite reduction zone and hence it would result in carbon and blast requirements smaller than those predicted by equations (7) and (8).

Metallic iron does not, however, seem to be present in significant quantities in the Fe_2O_3, Fe_3O_4 reduction regions of the blast furnace (Kanfer, 1974). This observation, coupled with the agreement between efficient industrial practice and the equation (7, 8) predictions (Chapter 13), suggests that the assumption of no iron reduction in the top segment of the conceptually divided furnace is valid.*

10.5.2 Magnetite Descent into the Wustite Reduction Zone

The gases rising from the wustite/Fe reduction zone contain more than enough CO to complete Fe_2O_3/Fe_3O_4/wustite reduction high in the

---

*In graphical terms this conclusion is equivalent to acknowledging that the blast-furnace operating line (Sections 4.5, 9.3) cannot lie to the right of graphical point W (Figs. 9.1, 10.1).

stack, Section 2.7. There is the possibility, however, that the occasional piece of Fe_3O_4 might, for kinetic reasons, descend unreduced through the chemical reserve into the wustite reduction zone. This would mean that some of the CO which could have reduced wustite to Fe would be wasted in reducing Fe_3O_4 to $Fe_{0.947}O$. The net results would be that (i) an increased amount of CO would have to be produced, (ii) carbon consumption would be increased. Fe_3O_4 descent into the wustite reduction zone would, therefore, result in higher carbon consumptions than those predicted by equations (7) and (8). However, blast furnaces seem to have evolved to a sufficient height ($\simeq$30 m) to ensure magnetite reduction high in the shaft. The assumption that magnetite does not descend into the wustite reduction zone appears, therefore, to be justified.

10.5.3 Carbon Oxidation Above the Wustite Reduction Zone

Figure 3.4 shows that the rate of carbon oxidation by CO_2 decreases by almost two orders of magnitude for each 100 K drop in temperature. It is unlikely, therefore, that a significant quantity of carbon will be oxidized in the high, cool parts of the shaft, i.e. in the top segment of the conceptually divided furnace. Thus the predictions of equations (7) and (8) are not likely to be invalidated by carbon oxidation above the wustite reduction zone.

10.5.4 Discussion

Although Fe_3O_4 or Fe descent into the wustite reduction zone and carbon oxidation in the top segment of the furnace are all possible, they do not seem to significantly affect the overall blast-furnace process. It is interesting to note further that:

(a) Fe_3O_4 reduction in the wustite reduction zone and carbon oxidation high in the shaft would both result in consumptions of carbon higher than those predicted by equations (7) and (8).
(b) Fe production high in the stack would have the opposite effect.

Thus, even if these reactions were significant, they would tend to offset each other to some extent.

10.6 Non-attainment of Equilibrium in the Chemical Reserve Zone

Equations (7) and (8) assume that Reaction (1.3),

$$CO + Fe_{0.947}O \rightleftharpoons 0.947Fe + CO_2, \qquad (1.3)$$

approaches equilibrium in a chemical reserve. There is always the possibility, however, that the ascending gas might rise so quickly through the chemical reserve that it would be unable to come near its equilibrium composition. In this case, the CO_2/CO ratio of the gas leaving the reserve would be lower than the value of 0.3/0.7 used in developing the model, i.e. the actual value of $(O/C)^{gwrz}$ would be lower than the model value, $(O/C)^{gwrz} = 1.3$. The net result would (equation (9a)) be an actual carbon requirement somewhat higher than that predicted by the model.

Physically, a departure from the Reaction (1.3) equilibrium is indicative of an inefficient use of reducing gas in the wustite reduction zone, i.e. it indicates that the reducing gas is not removing its full capacity of oxygen from $Fe_{0.947}O$. This situation might, for example, arise if (i) a furnace is being pushed (by a high rate of blast and with an excessive gas velocity) for a high production rate or if (ii) the gas is channeling through open spaces in the charge. In either case the net result would be an increased requirement for carbon.

Figure 10.1 shows graphically the effect of the gas not coming close to the $Fe_{0.947}O/Fe$ equilibrium. The heat demand of the process (i.e. the position of point H) remains unchanged, but point W (Section 9.4.1) moves to the left. The net results are: (i) an increased slope of the operating line which is equivalent to an increase in the carbon requirement and (ii) an increased blast oxygen requirement n_O^B (as indicated by a more negative intercept at O/C = 0). Thus the model clearly shows that pushing for a high production rate at the expense of having the CO remove its full capacity of oxygen from $Fe_{0.947}O$ in the wustite reduction zone raises the carbon and blast requirements of the process.

10.7 Thermal Reserve Temperature Effects

To this point it has been assumed that the thermal reserve temperature is always 1200 K. In this regard, it will be remembered that the thermal reserve temperature is that at which the rate of coke gasification becomes

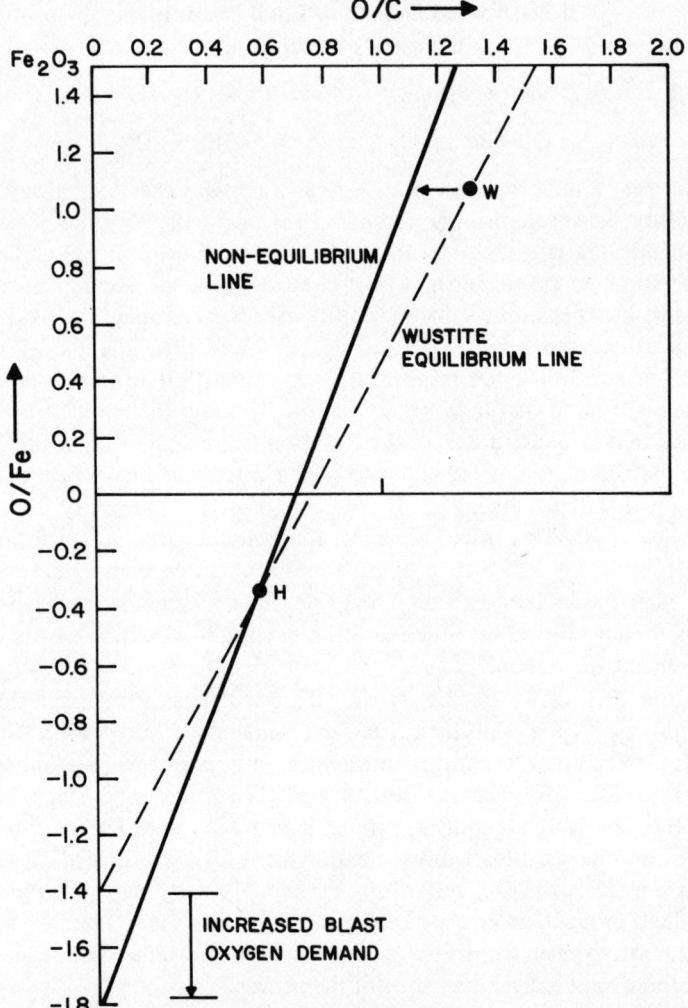

Fig. 10.1. Blast furnace *operating lines* when (i) $CO/CO_2/Fe/Fe_{0.947}O$ equilibrium is achieved in the chemical reserve (----) and when (ii) equilibrium is not attained in the chemical reserve (———). The two lines are for the same operation, i.e. for an identical heat demand. The steep slope and deep intercept of the non-equilibrium line are indicative of large carbon and blast demands.

Testing of the Mathematical Model

negligible. The actual temperature at which this occurs depends upon coke reactivity, and for this reason the thermal reserve temperature may vary slightly from furnace to furnace. It can be reasonably suggested, however, that the thermal reserve temperatures of industrial furnaces will be within the range 1100 to 1300 K.

The temperature of the thermal reserve affects:*

(a) the heat demand of the wustite reduction zone D^{wrz} which tends to be high when the reserve temperature is low, because the charge descending into the wustite reduction zone must be heated to its final temperature (1800 K) from a cooler starting point;

(b) the gas composition in the chemical reserve zone $(O/C)^{gwrz}$, which decreases with increasing temperature, Fig. 3.3;

and thus it influences the carbon and blast requirements of the process (equations (7.1) and (8.3)).

It can be seen from the above discussion that predictions based on a 1200 K reserve temperature will be in error if the actual reserve temperature is 1100 or 1300 K. The magnitude of these potential errors is indicated in Table 10.2, which presents predictions of carbon and blast requirements based on thermal reserve temperatures of 1100, 1200 and 1300 K.

It can be seen from Table 10.2 that the above-mentioned ±100 K thermal reserve temperature uncertainty leads to an uncertainty in predicted carbon demand of the order of ±5%. The relatively small size of this uncertainty is due to:

(a) the small effect of temperature on $(O/C)^{gwrz}$, i.e. on the equilibrium CO/CO_2 ratio for wustite reduction (Fig. 3.3);

(b) the small effect of temperature on the wustite reduction zone heat demand D^{wrz}, due mainly to the temperature insensitivity of $H^f_{Fe_{0.947}O}$ (Appendix V).

In addition, the variations of D^{wrz} and $(O/C)^{gwrz}$ with temperature have offsetting effects on n^A_C, equation **(9a)**.

A thermal reserve temperature of 1200 K is assumed throughout this

---

*H^f_{CO} and $H^f_{CO_2}$ are almost unaffected (Appendix V), hence the enthalpy supply terms in equation (8) remain virtually constant.

TABLE 10.2.

Effects of Thermal Reserve Temperature on D^{wrz}, $(O/C)^{gwrz}$ and the Predictions of the Operating Equations.

All calculations assume minimum heat demand conditions (production of Fe[5% C] from wustite; no flux, gangue or heat losses); and a blast temperature of 1400 K. Equations (7.1) and (8.3) were used for the calculations.

	Thermal reserve temperature (K)		
	1100	1200 (Section 9.1)	1300
Parameter			
D^{wrz} (kJ kg mole Fe)$^{-1}$	335 000	330 000	326 000
$(O/C)^{gwrz}$	1.34	1.30	1.28
n_O^B	0.94	0.99	1.05
n_C^A	1.49	1.58	1.65
Total carbon			
$(n_C^A + 0.25)$ (kg moles of C per kg mole of Fe)	1.74	1.83	1.90

text but this uncertainty of ±5% must be kept in mind when the predictions are being evaluated.

10.8 Summary

This chapter has shown that the parameters calculated by means of wustite reduction zone equations (7) and (8) satisfy all temperature, stoichiometric and thermodynamic requirements of the blast furnace process.

The basic assumptions of the model, i.e. no descent of Fe or Fe_3O_4 into the wustite reduction zone and no oxidation of solid carbon high in the furnace, have been examined. These phenomena are all theoretically possible but the success of the model indicates that they do not significantly affect the overall behaviour of the furnace.

The effects of the lack of certainty with which the thermal reserve temperature is known have been evaluated. The uncertainty in carbon

Testing of the Mathematical Model

requirement is shown to be in the order of ±5% when the thermal reserve temperature uncertainty is ±100 K.

Significant departures from $CO/CO_2/Fe_{0.947}O/Fe$ equilibrium in the chemical reserve are shown to result in increased carbon and blast requirements for the process. This situation will arise if the furnace is pushed hard for a high production rate or if the reducing gases channel unreacted through the wustite reduction zone.

References

Kanfer, V. D. and Muravev, V. N. (1974) 'Investigation of production processes in a quenched industrial blast furnace', *Steel in the U.S.S.R.* 4(9), 864–868.

Kitaev, B. E., Yaroshenko, Y. U. and Suchkov, V. D. (1967) *Heat Exchange in Shaft Furnaces*, Pergamon Press, Oxford, pp. 183–186.

Stull, D. R., Prophet, H. *et al.* (1970) *JANAF Thermochemical Tables*, 2nd edition United States Department of Commerce, Document NSRDS-NBS 37, Washington, June 1971.

Chapter 10 Problems. *Top Gas Temperature, Departure from Equilibrium in the Chemical Reserve*

10.1 Calculate the temperature of the top gas leaving the blast furnace described in Problem 9.1.

10.2 The coke portion of the blast-furnace charge is often moist due to atmospheric precipitation and to water from coke quenching. To examine the effect of this moisture, recalculate Problem 10.1 for the case of coke containing 5 wt.% H_2O. (Assume that the moisture does not take part in any chemical reactions, i.e. that it merely vaporizes and warms.)

10.3 The blast furnace described in Problem 9.1 is 'pushed' for production by increasing its blast rate (Nm^3 per minute) and by increasing the rate of charging input solids. It is found that the first incremental changes in blast rate lead to proportionate increases in Fe production rate but that further changes are not commensurately rewarded. In addition, large increases in blast and charge rates result in increased blast and coke requirements. It is observed, for example, that the carbon requirement (per tonne of Fe) increases 10% when the blast rate (Nm^3 per minute) is increased 25%.

To evaluate these effects, calculate for the 'pushed' Problem 9.1 furnace:
(a) the blast requirement (per tonne of Fe);
(b) the carbon requirement (per tonne of Fe);
(c) the O/C ratio of the gas just as it first meets Fe_3O_4 (i.e. just as it leaves the wustite reduction zone).

Calculate also (d) the percentage by which the iron production rate has been increased by 'forcing' the furnace. Assume that the Fe production rate is proportional to the ratio: [blast rate (Nm^3 per minute)]/[blast requirement (Nm^3 per tonne of Fe)].

It should be remembered that in this problem, the wustite reduction zone heat demand (D^{wrz}) is unaffected by the quantities of coke and blast which are added to the furnace.

CHAPTER 11

The Effects of Tuyère Injectants on Blast-Furnace Operations

Up to this point, the basic mathematical model has assumed (i) that the source of all fuel and reducing gas is coke; and (ii) that the only material passing through the tuyères is air. The three basic blast-furnace-operating equations under these conditions are:*

(a) $n_O^B + (O/Fe)^x = n_C^A \cdot (O/C)^g$ (4)

(stoichiometric equation for the whole furnace);

(b) $n_O^B + 1.06 = n_C^A \cdot (1.3)$ (7)

(stoichiometric equation for the bottom segment of a conceptually divided furnace);

(c) $D^{wrz} = S^{wrz}$
$= n_C^A \cdot (198\,000) + E^B \cdot n_O^B$ (8)

(enthalpy equation for the bottom segment).

Equations (7) and (8) have been combined in Chapter 9 to provide a convenient equation for the *a priori* prediction of blast-furnace-operating requirements. For simplicity, however, they will be treated as separate equations in the early parts of this chapter.

This chapter considers (i) how the above equations must be modified to

*Where:

$1.06 = (O/Fe)^{xwrz}$ (wustite),

$1.3 = (O/C)^{gwrz}$ ($CO/CO_2/Fe_{0.947}O$ equilibrium at 1200 K),

$198\,000 = -\left[\left\{2 - (O/C)^{gwrz}\right\} \cdot H^f_{1200}{}_{CO} + \left\{(O/C)^{gwrz} - 1\right\} \cdot H^f_{1200}{}_{CO_2}\right]$.

account for materials injected into the blast furnace through the tuyères and (ii) how the tuyère-injected materials affect the overall blast-furnace operation. The injectants considered are:

(a) oxygen (i.e. oxygen enrichment of air blast);
(b) hydrocarbons (gas, liquid or solid);
(c) moisture (H_2O);
(d) mixtures of carbonaceous and hydrogenous gases.

These injectants are added to replace coke with a cheaper form of reductant and/or to increase furnace productivity.

Details of industrial injection techniques and their operational consequences can be obtained from publications listed at the end of this chapter.

11.1 A General Injectant*

The materials injected through blast-furnace tuyères always contain one or more of C, H_2 or O. A general injectant is, therefore, represented by the formula:

$$C_x(H_2)_y O_z.$$

Examples of x, y and z are:

CH_4 $x = 1; y = 2; z = 0,$
H_2O $x = 0; y = 1; z = 1,$
O_2 $x = 0; y = 0; z = 2.$

Hydrocarbon injectants are usually added to the furnace directly through the tuyères and hence they enter the furnace at room temperature ($\simeq$298 K). Pure oxygen is, on the other hand, added to the blast air prior to heating in the hot-blast stoves so that it enters the furnace at the temperature of the blast, as does the humidity in the ambient air. For simplicity, an injectant temperature of 298 K is assumed while developing the equations of this chapter, but actual injectant temperatures are readily inserted into the equations (footnote, page 132).

---

*Methods of representing complex injectants whose chemical analyses and heat of combustion are known but not their chemical structure (e.g. coal, petroleum industrial natural gas) are presented in Appendix II.

11.1.1 Individual Components of the Injectant

From a mass balance or stoichiometric point of view, it is useful to consider the individual amounts of C, H_2 and O which are brought into the furnace by an injectant. In terms of a specified amount of injectant, i.e.

$$n^I \text{ (moles of } C_x(H_2)_y O_z \text{ per mole of product Fe),}$$

these are

carbon: $x \cdot n^I$ (moles of C per mole of product Fe),
hydrogen: $y \cdot n^I$ (moles of H_2 per mole of product Fe),
oxygen: $z \cdot n^I$ (moles of O per mole of product Fe).

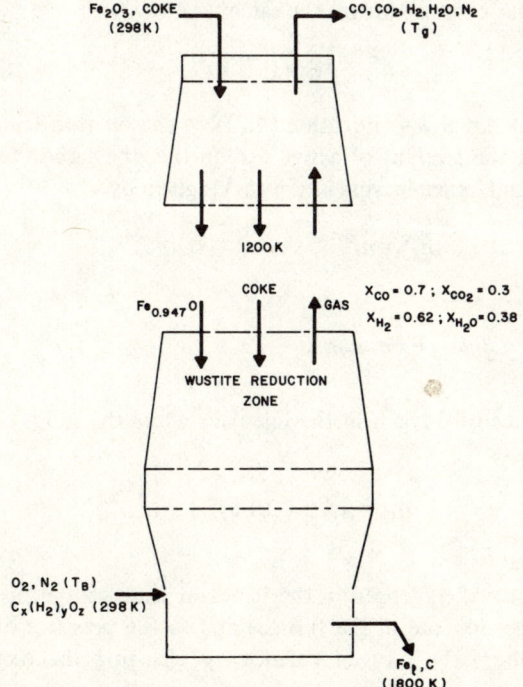

Fig. 11.1. Inputs and outputs of the top and bottom segments of a conceptually divided blast furnace when $C_x(H_2)_y O_z$ (a general tuyère injectant) is being injected through the tuyères.

11.2 Representing Injected Materials in the Overall Stoichiometric Equation

Figure 11.1 shows the input and output materials of a blast furnace into which $C_x(H_2)_y O_z$ is being injected. The presence of the injectant requires several modifications to the stoichiometric balances.

11.2.1 Carbon Balance

The injection of n^I moles of $C_x(H_2)_y O_z$ alters the carbon input to:

$$n_C^i = n_C^{coke} + x \cdot n^I$$

while the output carbon remains the same

$$n_C^o = n_C^A + (C/Fe)^m. \qquad \text{(Section 4.3)}$$

Since n_C^o must equal n_C^i (Equation (4.2)), it can be seen from the above equations that the amount of active carbon (i.e. the carbon taking part in the reactions and hence leaving in the gas) is given by

$$n_C^A = n_C^{coke} + x \cdot n^I - (C/Fe)^m. \qquad (11.1)$$

11.2.2 Oxygen Balance Equations

(a) *Input*

The presence of oxygen in the injectant alters the input oxygen equation to

$$n_O^i = n_O^B + (O/Fe)^x + z \cdot n^I. \qquad (11.2)$$

(b) *Output*

The presence of hydrogen in the injectant leads to hydrogen reduction of some of the iron ore in the furnace and to the presence of H_2O in the top gas. It alters the oxygen balance by changing the oxygen output equation to

$$n_O^o = n_{CO}^g + 2n_{CO_2}^g + n_{H_2O}^g. \qquad (11.3)$$

11.2.3 Development of the Stoichiometric Equation

As was shown in Chapter 4, the stoichiometric equation for the whole furnace is based primarily on the input—output oxygen balance. It is convenient, therefore, to put equations (11.2) and (11.3) into a form similar to that used in previous chapters. This is done by inserting, wherever possible, the terms:

n_C^A moles of carbon reacting with oxides and blast components (per mole of product Fe);

n^I moles of injectant (per mole of product Fe);

$X_{CO}^g, X_{CO_2}^g$ mole fractions of CO, CO_2 in the *carbonaceous portion* of the top gas, i.e.

$$X_{CO}^g = \frac{n_{CO}^g}{n_{(CO + CO_2)}^g} ; \quad X_{CO_2}^g = \frac{n_{CO_2}^g}{n_{(CO + CO_2)}^g}.$$

$X_{H_2}^g, X_{H_2O}^g$ mole fractions of H_2, H_2O in the *hydrogenous portion* of the top gas, i.e.

$$X_{H_2}^g = \frac{n_{H_2}^g}{n_{(H_2 + H_2O)}^g} ; \quad X_{H_2O}^g = \frac{n_{H_2O}^g}{n_{(H_2 + H_2O)}^g}.$$

As a first step in incorporating these terms, it can be noted that all the hydrogen entering in the injectant leaves the furnace in the top gas (i.e. very little leaves in the slag or metal), in which case

$$n_{(H_2 + H_2O)}^g = y \cdot n^I$$

and

$$n_{H_2}^g = y \cdot n^I \cdot X_{H_2}^g \quad \text{(from the definitions of } X_{H_2}^g, X_{H_2O}^g\text{)}$$

$$n_{H_2O}^g = y \cdot n^I \cdot X_{H_2O}^g.$$

Likewise, all the active carbon leaves in the furnace gases as CO or CO_2 (Section 4.1), hence:

$$n_{CO}^g = n_C^A \cdot X_{CO}^g,$$

$$n_{CO_2}^g = n_C^A \cdot X_{CO_2}^g,$$

all of which lead to a final, convenient form of oxygen output equation (11.3):

$$n_O^o = n_C^A \cdot (X_{CO} + 2X_{CO_2}) + y \cdot n^I \cdot X_{H_2O}^g. \tag{11.4}$$

The stoichiometric equation for the overall furnace is obtained by balancing the input and output oxygen expressions (i.e. $n_O^i = n_O^o$, equations (11.2), (11.4)), which gives

$$n_O^B + (O/Fe)^x + z \cdot n^I = n_C^A \cdot (X_{CO}^g + 2X_{CO_2}^g) + y \cdot n^I \cdot X_{H_2O}^g \tag{11.5}$$

where n_C^A, it must be remembered, is given by

$$n_C^A = n_C^{coke} + x \cdot n^I - (C/Fe)^m. \tag{11.1}$$

Equation (11.5) looks somewhat different than equation (4) (the stoichiometric equation without injection), but when there is no injection, i.e. when $n^I = O$, the equations are identical.*

11.3 Representing Injected Materials in the Bottom Segment Stoichiometric Equation

By similar reasoning to that of Section 11.2 above and Chapter 7, the stoichiometric equation for the conceptual bottom segment of the furnace is

$$n_O^B + (O/Fe)^{xwrz} + z \cdot n^I = n_C^A \cdot (X_{CO}^{gwrz} + 2X_{CO_2}^{gwrz}) + y \cdot n^I \cdot X_{H_2O}^{gwrz}$$

(11.7)

where

$(O/Fe)^{xwrz} = 1.06$ (i.e. the only iron oxide descending into the bottom segment is $Fe_{0.947}O$, Section 7.3),

$X_{CO}^{gwrz} = 0.7$ (i.e. the carbonaceous gases approach equilibrium with $Fe_{0.947}O$ and Fe, 1200 K, Section 7.3).

$X_{CO_2}^{gwrz} = 0.3$

*The terms X_{CO}^g, $X_{CO_2}^g$ are used here rather than $(O/C)^g$ so as not to confuse oxygen in the hydrogenous gases with oxygen in the carbonaceous gases. It will be remembered (Section 4.1) that $X_{CO}^g + 2X_{CO_2}^g = (O/C)^g$, where $(O/C)^g$ is the molecular oxygen to carbon ratio in the carbonaceous portion of the gas.

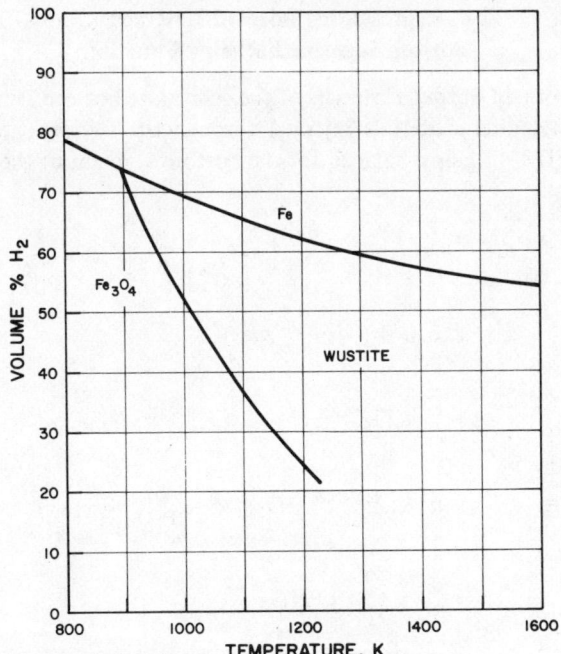

Fig. 11.2. Equilibrium gas compositions for the reaction

$$H_2 + Fe_{0.947}O \rightleftharpoons 0.947Fe + H_2O_{(g)}$$

expressed as volume% H_2 in the hydrogenous portion of the gas (thermodynamic data: Stull, 1970).

H_2 and H_2O, like the carbonaceous gases, also approach equilibrium with $Fe_{0.947}O$ and Fe in the chemical reserve, in which case (Fig. 11.2)

$$X_{H_2}^{gwrz} = 0.62,$$

$$X_{H_2O}^{gwrz} = 0.38.$$

Inserting these terms, the stoichiometric equation for the bottom segment (equation (11.7)) becomes:

$$n_O^B + z \cdot n^I + 1.06 = n_C^A \cdot (1.3) + y \cdot n^I \cdot (0.38). \tag{11.8}$$

11.4 Representing Injected Materials in the Bottom Segment Enthalpy Equation

The input and output materials of the schematic bottom segment of the furnace, including materials injected through the tuyères, are shown in Fig. 11.1. The enthalpy balance for the bottom segment of the furnace on this basis is:

$$1.06 \cdot H^\circ_{1200} \underset{Fe_{0.947}O}{} + n^{coke}_C \cdot H^\circ_{1200} \underset{C}{} + n^B_{O_2} \cdot \tfrac{1}{2} \cdot H^\circ_{T_B} \underset{O_2}{} + \frac{0.79}{0.21} \cdot \tfrac{1}{2} \cdot H^\circ_{T_B} \underset{N_2}{} + n^I \cdot H^\circ_{298} \underset{C_x(H_2 \ldots}{}$$

$$= H^\circ_{1800} \underset{Fe_l}{} + (C/Fe)^m \cdot H_{1800} \underset{C}{}$$

$$+ n^A_C \cdot (X^{gwrz}_{CO} \cdot H^\circ_{1200} \underset{CO}{} + X^{gwrz}_{CO_2} \cdot H^\circ_{1200} \underset{CO_2}{})$$

$$+ y \cdot n^I \cdot (X^{gwrz}_{H_2} \cdot H^\circ_{1200} \underset{H_2}{} + X^{gwrz}_{H_2O} \cdot H^\circ_{1200} \underset{H_2O}{})$$

$$+ n^B_{O_2} \cdot \frac{0.79}{0.21} \cdot \tfrac{1}{2} \cdot H^\circ_{1200} \underset{N_2}{} . \tag{11.9}$$

This equation, without tuyère injectants, is equivalent to equation (8.1). To avoid confusion, the number of moles of CO and CO_2 leaving the bottom segment are expressed in terms of X^{gwrz}_{CO} and $X^{gwrz}_{CO_2}$ rather than $(O/C)^{gwrz}$, but otherwise the equations are similar.

Equation (11.9) is put into a more useful form by making the substitutions

$$1.06 \cdot H^\circ_{1200} \underset{Fe_{0.947}O}{} = 1.06 \cdot H^f_{1200} \underset{Fe_{0.947}O}{} + H^\circ_{1200} \underset{Fe_s}{} + 1.06 \cdot \tfrac{1}{2} \cdot H^\circ_{1200} \underset{O_2}{},$$

$$H^\circ_{1200} \underset{CO}{} = H^f_{1200} \underset{CO}{} + H^\circ_{1200} \underset{C}{} + \tfrac{1}{2} \cdot H^\circ_{1200} \underset{O_2}{},$$

$$H^\circ_{1200} \underset{CO_2}{} = H^f_{1200} \underset{CO_2}{} + H^\circ_{1200} \underset{C}{} + H^\circ_{1200} \underset{O_2}{},$$

$$H^\circ_{1200} \underset{H_2O}{} = H^f_{1200} \underset{H_2O}{} + H^\circ_{1200} \underset{H_2}{} + \tfrac{1}{2} \cdot H^\circ_{1200} \underset{O_2}{},$$

$$H^\circ_{298} \atop C_x(H_2)O_z} = {H^f_{298} \atop C_x(H_2)_yO_z} + x \cdot {H^\circ_{1200} \atop C} + y \cdot {H^\circ_{1200} \atop H_2} + z \cdot \tfrac{1}{2} \cdot {H^\circ_{1200} \atop O_2}$$

$$- x \cdot [H^\circ_{1200} - H^\circ_{298}]_C - y \cdot [H^\circ_{1200} - H^\circ_{298}]_{H_2}$$

$$- z \cdot \tfrac{1}{2} \cdot [H^\circ_{1200} - H^\circ_{298}]_{O_2}$$

and

$$n^A_C = n^{coke}_C + x \cdot n^I - (C/Fe)^m. \qquad (11.1)$$

Making these substitutions; rearranging; and changing all the signs to put the equation in supply/demand form, equation (11.9) becomes:

$$-1.06 \cdot {H^\circ_{1200} \atop Fe_{0.947}O} + {[H^\circ_{1800} - H^\circ_{1200}] \atop Fe_l} + {(C/Fe)^m \cdot [H_{1800} - H^\circ_{1200}] \atop Fe_s} \quad {[C] \atop C}$$

$$+ n^I \cdot \left\{ -{H^f_{298} \atop C_x(H_2)_yO_z} + x \cdot [H^\circ_{1200} - H^\circ_{298}]_C + y \cdot [H^\circ_{1200} - H^\circ_{298}]_{H_2} \right.$$

$$\left. + z \cdot \tfrac{1}{2} \cdot [H^\circ_{1200} - H^\circ_{298}]_{O_2} \right\}$$

$$= -n^A_C \cdot (X^{gwrz}_{CO} \cdot {H^f_{1200} \atop CO} + X^{gwrz}_{CO_2} \cdot {H^f_{1200} \atop CO_2}) - y \cdot n^I \cdot X^{gwrz}_{H_2O} \cdot {H^f_{1200} \atop H_2O} + E^B \cdot n^B_O$$

$$- \tfrac{1}{2} \cdot {H^\circ_{1200} \atop O_2} \cdot \left\{ n^A_C \cdot (X^{gwrz}_{CO} + 2X^{gwrz}_{CO_2}) + y \cdot n^I \cdot X^{gwrz}_{H_2O} - z \cdot n^I - n^B_O - 1.06 \right\}.$$

$$(11.10)$$

The last term of this equation is zero (from equation (11.7)) which leads, for the conditions cited in Section 11.3 and Fig. 11.1, to the supply–demand equation:

$$D^{wrz} + n^I \cdot D^I = S^{wrz}$$
(wustite reduction zone heat supply)

$$= -n^A_C \cdot (0.7 \cdot {H^f_{1200} \atop CO} + 0.3 \cdot {H^f_{1200} \atop CO_2}) - 0.38 \cdot y \cdot n^I \cdot {H^f_{1200} \atop H_2O} + E^B \cdot n^B_O.$$

$$(11.11)$$

The important derived terms in this equation are:

D^{wrz} the heat demand of the wustite reduction zone, which in the simplest case is the heat required to form liquid iron (5% C, 1800 K) from $Fe_{0.947}O$ (1200 K), kJ (kg mole of Fe)$^{-1}$;

E^B the enthalpy in the blast air above that which it would have at 1200 K, kJ (kg mole of O from dry blast air)$^{-1}$;

D^I the injectant heat demand, kJ (kg mole of injectant)$^{-1}$.

The injectant heat demand is a new term. It has the form

$$D^I = -H^f_{298\ C_x(H_2)_yO_z} + x \cdot [H^\circ_{1200} - H^\circ_{298}]_C + y \cdot [H^\circ_{1200} - H^\circ_{298}]_{H_2}$$
$$+ z \cdot \tfrac{1}{2} \cdot [H^\circ_{1200} - H^\circ_{298}]_{O_2}$$

and it is the heat required to form molecular C, H_2 and O_2 (1200 K) from $C_x(H_2)_yO_z$ (298 K), kJ per kg mole of injectant.* Of course, if molecular C, H_2 or O_2 are the injectants,

$$H^f_{298\ C_x(H_2)_yO_z} = 0$$

and $$D^I = [H^\circ_{1200} - H^\circ_{298}]_{C,H_2,O_2}$$

Lastly it can be seen that when $n^I = 0$, i.e. when there are no injectants, equation (11.11) simplifies to equation (8).

---

*Heat demands of injectants entering at temperatures other than 298 K, e.g. water vapour and pure oxygen, are represented by

$$D^I_{(T)} = -H^f_{T\ C_x(H_2)_yO_z} + x \cdot [H^\circ_{1200} - H^\circ_T]_C + y \cdot [H^\circ_{1200} - H^\circ_T]_{H_2}$$
$$+ z \cdot \tfrac{1}{2} \cdot [H^\circ_{1200} - H^\circ_T]_{O_2}.$$

In the absence of other information, H^f_T for complex injectants (e.g. coal, oil, Appendix II) can be assumed to equal H^f_{298}.

11.5 A Form Convenient for Calculations

Equations (11.8) and (11.11) could be combined in a like manner to that used in developing equation (9) (stoichiometry/enthalpy equation for the wustite reduction zone) but this would result in a rather complicated final equation. It appears to be more straightforward to treat equations (11.1), (11.8) and (11.11) by simultaneous methods.

Thus the only further step in improving the equations is to make the numerical substitutions

$$H^f_{1200}_{CO} = -113\,000,$$

$$H^f_{1200}_{CO_2} = -395\,000 \text{ kJ} \qquad \text{(kg mole of gas)}^{-1},$$

$$H^f_{1200}_{H_2O} = -249\,000$$

into equation (11.11), which becomes:

$$D^{wrz} + n^I \cdot D^I = S^{wrz}$$

$$= -n^A_C \cdot \left\{0.7 \cdot (-113\,000) + 0.3 \cdot (-395\,000)\right\}$$

$$-0.38 \cdot (-249\,000) \cdot y \cdot n^I + E^B \cdot n^B_O$$

$$= n^A_C \cdot \left\{198\,000\right\} + y \cdot n^I \cdot (95\,000) + E^B \cdot n^B_O. \qquad (11.12)$$

11.6 Example Calculations: I. Oxygen Enrichment

The addition of pure oxygen to the blast air (i.e. oxygen enrichment or injection)

(a) raises the flame temperature in the tuyère zone;
(b) lowers the volume of gas which must be blown through the tuyères per unit of product Fe;
(c) lowers the volume of gas which must rise through the furnace per unit of product Fe.

The Iron Blast Furnace

All of these effects are due to less nitrogen being blown into the furnace (per unit of oxygen entering the furnace) and they can all be beneficial to blast-furnace operation. The higher flame temperature permits injection of cold hydrocarbons through the tuyères while still maintaining an acceptable hearth temperature; the lower volume of gas through the tuyères permits a higher iron production rate for a given blower capacity; and the lower volume of gas rising through the furnace shaft permits a higher iron production rate without 'flooding' the bosh or causing the ascending gases to 'channel' through the charge.

As an illustrative problem, consider the changes which would arise if 50 Nm^3 of O_2 per tonne of Fe were added to the blast air (prior to preheating) of the furnace described in the illustrative problem of Section 9.4. The specifications for the oxygen-enriched operation are seen in Table 11.1.

The injectant heat demand in this example (for O_2 at 1400 K) is given (footnote, page 132) by

$$D^I = [H°_{1200} - H°_{1400}]_{O_2}$$

$$= -7000 \text{ kJ (kg mole of injectant)}^{-1}$$

so that equations (11.12) and (11.8) are:

$$400\,000 + 0.125 \cdot (-7000) = n^A_C \cdot (198\,000) + 17\,000 \cdot n^B_O, \quad (11.12)$$

$$n^B_O + 2 \cdot (0.125) + 1.06 = n^A_C \cdot (1.3). \quad (11.8)$$

These two simultaneous equations are readily solved to give

$n^A_C = 1.91$ moles of C per mole of product Fe,

$n^B_O = 1.18$ moles of O per mole of product Fe,

from which the operating requirements for the operation are shown in Table 11.2.

11.6.1 Discussion of Oxygen Injection

Comparison of the data in Sections 9.4 and 11.6 shows that the addition of pure oxygen (1400 K with the blast air) to the furnace causes a

TABLE 11.1

Item	Specification	Quantity		Model variable (kg moles per kg mole of product Fe)	
		kg per tonne of Fe	kg moles per tonne of Fe		
Fe		1000	17.9		
Iron oxide entering wustite reduction zone	$Fe_{0.947}O$			$(O/Fe)^{xwrz}$	$= 1.06$
Pig iron	5% C			$(C/Fe)^m$	$= 0.25$
Blast temperature				T_B	$= 1400$ K
Blast enthalpy (Section 9.1)				E^B	$= 17\,000$ kJ per kg mole of O
Heat demand of wustite reduction zone (Section 9.4)				D^{wrz}	$= 400\,000$ kJ per kg mole of product Fe
Injectant	Pure oxygen enters furnace at blast temperature	50 Nm³	2.2	n^I $x = 0, y = 0, z = 2$ T	$= 0.125$ $= T_B$

136   *The Iron Blast Furnace*

TABLE 11.2

Item	Specification	Quantity		Model variable (kg moles per kg mole of product Fe)
		kg per tonne of Fe	kg moles per tonne of Fe	
Injectant	Pure oxygen (1400 K)	50 Nm³	2.2	$n^I = 0.125$ $z = 2$
Oxygen from dry blast air		338 kg (1130 Nm³ of air)	21.1 (O)	$n_O^B = 1.18$
Active carbon			34.2	$n_C^A = 1.91$
Total carbon		464	38.7	$n_C^t = (C/Fe)^m + n_C^A$ $= 0.25 + 1.91$ $= 2.16$

slight increase in the total carbon and total oxygen requirements of the process, i.e.

	Air blast only (Section 9.4) kg per tonne of Fe	Injection of 50 Nm^3 of pure O_2 (Section 11.6) kg per tonne of Fe
Total C requirement	462	464
Total O_2 requirement (pure O_2 + O_2 from air)	404	409

These increases in carbon and oxygen demand are due to a decrease in the amount of high temperature nitrogen entering the furnace. This has the net effect of decreasing the high temperature (>1200 K) enthalpy supply to the wustite reduction zone and it necessitates the burning of more carbon.*

11.6.2 Lowered Gas Volumes Due to Oxygen Injection

Referring again to Sections 9.4 and 11.6, it can be seen that the volume of gas which must be blown through the tuyères per tonne of Fe is considerably lowered by the injection of pure oxygen, i.e.

Total gas volume through tuyères (per tonne of Fe)

without O_2 injection 1350 Nm^3 (air),
with O_2 injection 1180 Nm^3 (air plus oxygen).

*Oxygen enrichment always decreases the amount of nitrogen entering the furnace per kg mole of product Fe. This can have three effects on the bottom segment enthalpy balance depending on the temperature at which the air blast is entering the furnace.

(i) T_B > 1200 K (thermal reserve temperature): nitrogen supplies enthalpy to the wustite reduction zone, hence its removal requires a compensating enthalpy supply (i.e. more carbon and oxygen).
(ii) T_B = 1200 K: nitrogen is neutral (in the enthalpy sense) and it has no effect on the carbon and oxygen demands of the wustite reduction zone.
(iii) T_B < 1200 K: nitrogen requires heating in the wustite reduction zone (i.e. to the thermal reserve temperature) so that its removal actually lowers demands for carbon and oxygen.

This means that if blower capacity is the production-limiting parameter of the operation, oxygen injection can lead to a considerable increase in furnace productivity, in this example about 10%.

Likewise, the volume of gases rising through the furnace (per tonne of product Fe) is lowered which means that the production rate of iron can be increased without channeling and without fluidizing or flooding the charge.

11.6.3 Graphical Calculations (Oxygen Enrichment)

In a similar manner to that outlined in Section 9.3, equations (11.8) and (11.12) (oxygen injection: $x = 0$, $y = 0$, $z = 2$) can be combined and arranged to:

$$1.06 - \left\{ 1.06 - \frac{D^{wrz} + n^I \cdot D^I + 1.06 \cdot E^B + z \cdot n^I \cdot E^B}{282\,000 + E^B} \right\}$$

$$= n_C^A \cdot \left\{ 1.3 - \frac{169\,000}{282\,000 + E^B} \right\}$$

which shows that in this case the coordinates of points W and H on an O/C : O/Fe plot are:

point W $\quad$ O/C = 1.3; $\quad\quad\quad$ O/Fe = 1.06,

point H $\quad$ O/C = $\dfrac{169\,000}{282\,000 + E^B}$;

$$O/Fe = 1.06 - \frac{D^{wrz} + n^I \cdot D^I + 1.06 \cdot E^B + z \cdot n^I \cdot E^B}{282\,000 + E^B}$$

Similarly, the bottom segment stoichiometric equation (11.8) can be expressed (for oxygen enrichment) as

$$1.06 - (-n_O^B - z \cdot n^I) = n_C^A \cdot (1.3 - 0)$$

which shows that the coordinates at the bottom of the operating line are:

$\quad$ O/C = 0; $\quad$ O/Fe = $-(n_O^B + z \cdot n^I)$

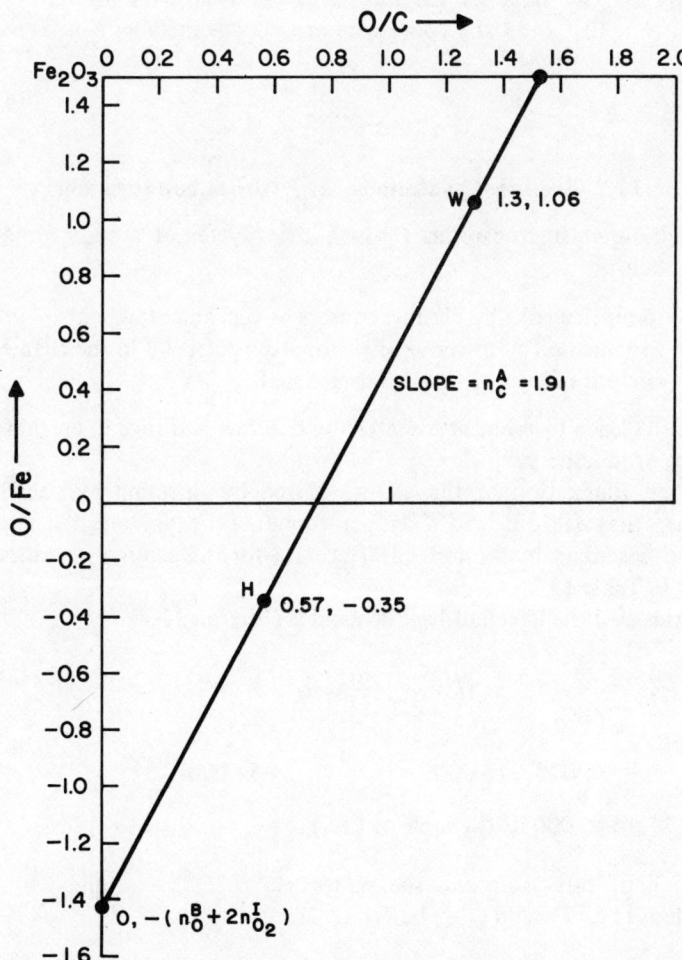

Fig. 11.3. Blast-furnace *operating line* when the blast air is being enriched with $n_{O_2}^I$ kg moles of pure O_2 per kg mole of Fe. For point H coordinates see Section 11.6.3. The furnace operation described by this diagram is the same as that in Fig. 9.1 but with a blast enrichment of 50 Nm^3 of pure O_2 per tonne of Fe.

The method of calculation (Fig. 11.3) is similar to that described in Section 9.4. The slope of the line is n_C^A as usual, and the only major difference is that the O/Fe coordinate at O/C = 0 includes $z \cdot n^I$ as well as n_O^B.

11.7 Example Calculations: II. Hydrocarbon Injection

Injection of hydrocarbons through the tuyères of a blast furnace is carried out

(a) to replace coke by cheaper sources of fuel and reductant;
(b) to increase (by lowering the proportion of coke in the charge) the amount of iron ore in the furnace shaft.

Effect (b) leads to a higher overall reduction rate and thus to an enhanced furnace productivity.

As an illustration of the effects caused by injecting hydrocarbons, consider that 100 Nm³ of CH_4 per tonne of Fe are injected into the furnace described in Section 9.4. The data for this altered operation are shown in Table 11.3.

In this case the injectant heat demand is (Section 11.4)

$$D^I = -H_{298}^f \big|_{CH_4} + 1 \cdot [H_{1200}^\circ - H_{298}^\circ]_C + 2 \cdot [H_{1200}^\circ - H_{298}^\circ]_{H_2}$$

$$= 75\,000 + 16\,000 \qquad\qquad + 54\,000$$

$$= 145\,000 \text{ kJ (kg mole of } CH_4)^{-1}.$$

Insertion of this value and the values $n^I = 0.25$; $x = 1$; and $y = 2$ into equations (11.12) and (11.8) leads to:

$$400\,000 + 0.25 \cdot (145\,000) = n_C^A \cdot (198\,000) + 2 \cdot (0.25) \cdot (95\,000)$$

$$+ 17\,000 \cdot n_O^B, \qquad (11.12)$$

$$n_O^B + 1.06 = n_C^A \cdot (1.3) + 2 \cdot (0.25) \cdot (0.38), \qquad (11.8)$$

TABLE 11.3

Item	Specification	Quantity		Model variable (kg moles per kg mole of product Fe)
		kg per tonne of Fe	kg moles per tonne of Fe	
Fe		1000	17.9	
Iron oxide entering wustite reduction zone	$Fe_{0.947}O$			$(O/Fe)^{x\,wrz} = 1.06$
Pig iron	5% C	53 (carbon)	4.4	$(C/Fe)^m = 0.25$
Blast temperature				$T_B = 1400\,K$
Blast enthalpy (Section 9.1)				$E^B = 17\,000$ kJ per kg mole of O
Heat demand of wustite reduction zone (Section 9.4)				$D^{wrz} = 400\,000$ kJ per kg mole of product Fe
Injectant (natural gas)	CH_4 (298 K)	100 Nm³ 71 kg	4.5	$n^I = 0.25$ $x = 1, y = 2, z = 0$

from which

$$n_C^A = 1.83 \text{ moles per mole of product Fe},$$

$$n_O^B = 1.51 \text{ moles of O per mole of product Fe}.$$

The active carbon can be further broken down into its sources by means of equation (11.1) ($x = 1$), i.e.

$$n_C^A = 1.83 = n_C^{coke} + x \cdot n^I - (C/Fe)^m$$

$$= n_C^{coke} + 1 \cdot (0.25) - 0.25$$

which leads to

$$n_C^{coke} = 1.83 \text{ (moles of C from coke)},$$

$$x \cdot n^I = 0.25 \text{ (moles of C from injectant)}$$

per mole of product Fe.

The final data in the completed table of operating parameters are shown in Table 11.4.

By reference to Section 9.4 it can be seen that 100 Nm³ of natural gas (71 kg) per tonne of product Fe lowers the coke requirement of the process by approximately 70 kg, i.e. the weight replacement ratio is of the order of 1 kg of coke per kg of CH_4. This lowered coke requirement produces the desired effect of freeing more of the blast furnace shaft for iron ore.

It can also be noticed that

(a) the total carbon requirement of the process is slightly decreased by injecting CH_4, i.e. from 462 kg of C (coke only) to 447 kg of C (coke + CH_4) per tonne of product Fe. This is due to the contribution of hydrogen towards reducing $Fe_{0.947}O$ in the wustite reduction zone;

(b) the blast oxygen requirement is slightly increased (in this case 404 kg to 433 kg per tonne of product Fe) when natural gas is injected. This extra oxygen is necessary to compensate (by combusting extra carbon) for the heat demand ($n^I D^I$) of the injectant. It is for this enthalpy reason that the hydrogen in the natural gas does not lead to a large carbon saving.

TABLE 11.4

Item	Specification	Quantity		Model variable (kg moles per kg mole of product Fe)
		kg per tonne of Fe	kg moles per tonne of Fe	
Oxygen from dry blast air		433 (1440 Nm³ of air)	27.1 (O)	$n_O^B = 1.51$
Carbon from coke		393	32.8	$n_C^{coke} = 1.83$
Carbon from injectant		54	4.5	$n_C^I = 0.25$
Total carbon		447	37.3	$n_C^t = 2.08$

144 *The Iron Blast Furnace*

11.8 Graphical Calculations (General Case)

The presence of hydrogen and water vapour in the furnace considerably complicates the stoichiometric diagram. However, the diagram still provides a visual concept of how the furnace must operate and it is a useful tool even in its more complicated form.

The source of the complication is that the terms

$$(O/C)^g, (O/C)^{gwrz},$$

taken at face value, would have to include the oxygen which is present in the water vapour. The simplest way to get around this problem is to define $(O/C)^g$ and $(O/C)^{gwrz}$ specifically as

(O/C) = ratio of oxygen to carbon *in the carbonaceous gases*.

In terms of mole fractions of CO and CO_2 in the carbonaceous gases, this ratio is given (Section 4.1.1) by

$$(O/C) = X_{CO} + 2X_{CO_2}.$$

A parallel term for the hydrogenous gases is (O/H_2) which is given by

$$(O/H_2) = X_{H_2O} \qquad \text{(Section 11.2)},$$

where X_{H_2O} is the mole fraction of H_2O in the hydrogenous gases (i.e. 1 mole of hydrogenous gas contains 1 mole of H_2 and X_{H_2O} moles of O).

11.8.1 Bottom Segment Stoichiometric Equation

In terms of the O/C and O/H_2 parameters, equation (11.7)

$$n_O^B + (O/Fe)^{xwrz} + z \cdot n^I = n_C^A \cdot (X_{CO}^{gwrz} + 2X_{CO_2}^{gwrz}) + y \cdot n^I \cdot X_{H_2O}^{gwrz}$$

becomes

$$n_O^B + (O/Fe)^{xwrz} + z \cdot n^I - y \cdot n^I \cdot (O/H_2)^{gwrz} = n_C^A \cdot (O/C)^{gwrz}. \tag{11.7a}$$

Further, when equilibrium conditions are approached in the chemical reserve:

$$(O/Fe)^{xwrz} = 1.06,$$

The Effects of Tuyère Injectants

$$(O/C)^{gwrz} = 1.3,$$
$$(O/H_2)^{gwrz} = 0.38,$$

from which:

$$1.06 - \left\{-n_O^B - z \cdot n^I + y \cdot n^I \cdot (0.38)\right\} = n_C^A \cdot (1.3 - 0).$$

As indicated in Section 9.4, this equation describes a straight line of slope n_C^A between the points:

$$O/C = 0; \qquad O/Fe = -\left\{n_O^B + z \cdot n^I - y \cdot n^I \cdot (0.38)\right\}$$

and

$$O/C = 1.3; \qquad O/Fe = 1.06 \qquad \text{(point W)}.$$

11.8.2 Bottom Segment Stoichiometry/Enthalpy Equation

Equations (11.7) and (11.11) may be combined and rewritten to give:

$$D^{wrz} + n^I D^I + z \cdot n^I \cdot E^B + 1.06 \cdot E^B - X_{H_2O}^{gwrz} \cdot y \cdot n^I \cdot (249\,000 + E^B)$$

$$= -n_C^A \cdot \left\{ X_{CO}^{gwrz} \cdot \underset{CO}{(H_{1200}^f)} + X_{CO_2}^{gwrz} \cdot \underset{CO_2}{(H_{1200}^f)} \right\}$$

$$+ (X_{CO}^{gwrz} + 2 X_{CO_2}^{gwrz}) \cdot E^B. \qquad (11.13)$$

Furthermore, since $(O/H_2)^{gwrz} = X_{H_2O}^{gwrz}$

and $\qquad (O/C)^{gwrz} = X_{CO}^{gwrz} + 2 X_{CO_2}^{gwrz}$

equation (11.13) becomes

$$D^{wrz} + n^I \cdot D^I + z \cdot n^I \cdot E^B + 1.06 \cdot E^B - (O/H_2)^{gwrz} \cdot y \cdot n^I \cdot (249\,000 + E^B)$$

$$= -n_C^A \cdot (O/C)^{gwrz} \cdot (H_{1200\,CO_2}^f - H_{1200\,CO}^f - E^B) + 2H_{1200\,CO}^f - H_{1200\,CO_2}^f,$$

(11.14)

or in numerical terms (i.e. from equations (11.8) and (11.12))

$$D^{wrz} + n^I \cdot D^I + z \cdot n^I \cdot E^B + 1.06 \cdot E^B - 0.38 \cdot y \cdot n^I \cdot (249\,000 + E^B)$$
$$= n_C^A \cdot \left\{ 1.3 \cdot (282\,000 + E^B) - 169\,000 \right\}.$$ (11.15)

For plotting, this equation rearranges to

$$1.06 - \left\{ 1.06 - \frac{D^{wrz} + n^I \cdot D^I + z \cdot n^I \cdot E^B + 1.06 \cdot E^B - 0.38 \cdot y \cdot n^I \cdot (249\,000 + E^B)}{282\,000 + E^B} \right\}$$

$$= n_C^A \cdot \left\{ 1.3 - \frac{169\,000}{282\,000 + E^B} \right\}$$

which is a straight line (Section 9.4.1) of slope n_C^A, between the points:

$$O/C = \frac{169\,000}{282\,000 + E^B};$$

$$O/Fe = \left\{ 1.06 - \frac{D^{wrz} + n^I \cdot D^I + z \cdot n^I \cdot E^B + 1.06 \cdot E^B - 0.38 \cdot y \cdot n^I \cdot (249\,000 + E^B)}{282\,000 + E^B} \right\}$$

point H,

$O/C = 1.3;$ $O/Fe = 1.06$ point W.

Of course when $n^I = 0$, this plot takes the simple form described in Section 9.3.

11.8.3 Method of Calculation

As is shown in Section 9.4, the method of graphical calculation is:

(a) plot points W and H;

(b) draw a straight line through points H and W, and extend it down to O/C = 0;
  (c) the slope of the line is n_C^A;
  (d) the O/Fe value at the O/C = 0 intercept is $-(n_O^B + z \cdot n^I - 0.38 \cdot y \cdot n^I)$.

The operating parameters of the Section 11.7 illustrative example have

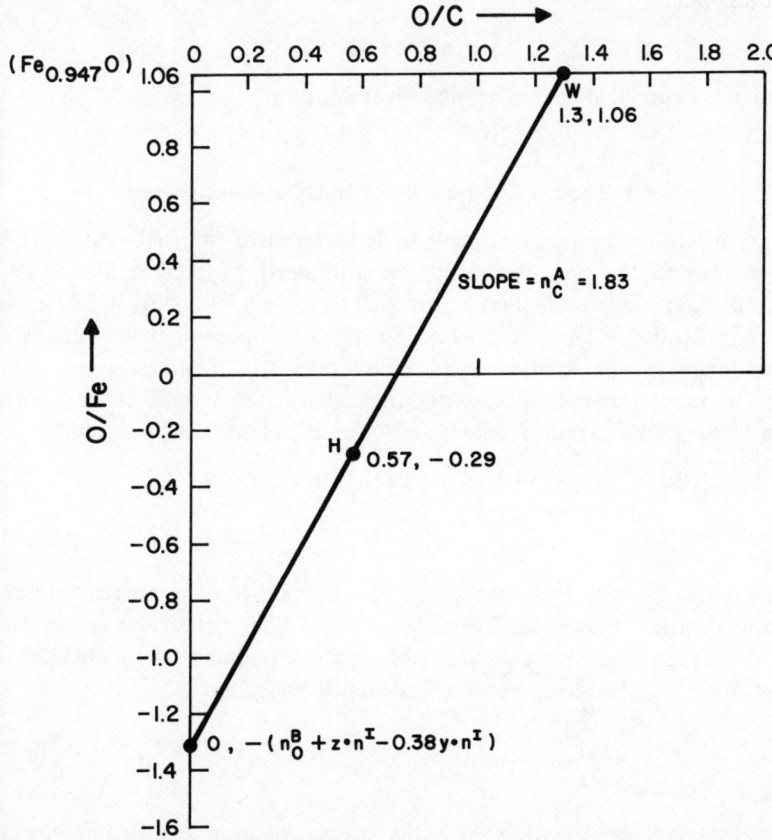

Fig. 11.4. Blast-furnace-*operating line* when n^I kg moles of $C_x(H_2)_y O_z$ are being injected into the furnace per kg mole of Fe. For point H coordinates see Section 11.8.2. The furnace operation described here is the same as that in Fig. 9.1 but with the injection of 100 Nm³ of CH_4 into the furnace per tonne of Fe.

been recalculated graphically as is shown in Fig. 11.4. The slope is

$$n_C^A = 1.83$$

which is, of course, identical to the numerical solution; while the intercept is:

$$-(n_O^B - 0.38 \cdot y \cdot n^I) = -1.32,$$

which, since $n^I = 0.25$, $y = 2$, $z = 0$, is equivalent to the blast oxygen requirement

$$n_O^B = 1.51,$$

which is again identical to the numerical solution.

11.9 Top-gas Composition with Hydrogen Injection

A blast-furnace operator may wish to estimate the fuel value of his blast-furnace top gas, for which he will need to predict his top-gas composition. This prediction is straightforward when the only reductant is carbon (Section 4.3), but it is somewhat more complicated when hydrogen (from H_2O in blast or tuyère injectants) is present in the furnace.

The stoichiometric equation for the top segment (Fig. 11.1) of a blast furnace into which hydrocarbons and/or moisture are being injected is

$$(O/Fe)^x - (O/Fe)^{xwrz} = n_C^A \cdot [(O/C)^g - (O/C)^{gwrz}]$$
$$+ n_{H_2}^I \cdot [(O/H_2)^g - (O/H_2)^{gwrz}], \quad (11.16)$$

where $n_{H_2}^I$ (active hydrogen) is the total quantity of hydrogen entering through the tuyères per kg mole of Fe. Under conditions where the carbonaceous and hydrogenous gases approach equilibrium with wustite and Fe in the chemical reserve, this equation becomes

$$(O/Fe)^x - 1.06 = n_C^A \cdot [(O/C)^g - 1.3] + n_{H_2}^I \cdot [(O/H_2)^g - 0.38].$$
$$(11.17)$$

Unfortunately equation (11.17) is a single equation with two unknowns which means that an assumption must be made as to the relative amounts of CO and H_2 taking part in higher oxide reduction. Perhaps the simplest assumption is that CO and H_2 remove oxygen from ore in the top segment in direct proportion to the quantity of each rising into that segment, i.e.

$$\frac{\text{kg moles of O picked up by CO in the top segment}}{\text{kg moles of O picked up by } H_2 \text{ in the top segment}} = \frac{n_{CO}^{gwrz}}{n_{H_2}^{gwrz}} \quad \text{(all per kg mole of Fe)}.$$

Expressed in model variables this equation is

$$\frac{n_C^A \cdot [(O/C)^g - 1.3]}{n_{H_2}^I \cdot [(O/H_2)^g - 0.38]} = \frac{n_C^A \cdot X_{CO}^{gwrz}}{n_{H_2}^I \cdot X_{H_2}^{gwrz}}$$

or (since $X_{CO}^{gwrz} = 0.7$; $X_{H_2}^{gwrz} = 0.62$ in the chemical reserve)

$$\frac{[(O/C)^g - 1.3]}{[(O/H_2)^g - 0.38]} = \frac{0.7}{0.62}. \tag{11.18}$$

Equations (11.17) and (11.18) are readily solved to give $(O/C)^g$ and $(O/H_2)^g$ from n_C^A and $n_{H_2}^I$. Complete analysis of the top gas can then be determined in a manner similar to that outlined in Section 4.3 (H_2O vapour from the solid charge, inactive hydrogen, must also be added in at this point).

11.10 Discussion of Injection Calculations and Summary

As previous sections of this chapter have shown, inclusion of the effects of tuyère injectants (in the general case $C_x(H_2)_y O_z$) does not change the form of the operating equations or the graphs. It simply increases the number of terms which must be taken into consideration.

One important point to note here is that in all the derivations of this chapter, the quantity of injectant is expressed per mole of product Fe. This is the simplest approach, but in many cases the quantity of injectant will be expressed in terms of the amount of blast which will require a change in the form of equation (11.12). This situation is posed as a problem (11.3) at the end of this chapter.

The situation of two injectants, for example CH_4 and pure O_2, is also not discussed in the text. The simplest way to handle this case is to consider the injectants as separate entities, i.e. in terms of separate quantities, heat demands and stoichiometric factors (x, y and z).

Reference

Stull, D. R., Prophet, H. *et al.* (1970) *JANAF Thermochemical Tables,* 2nd edition, United States Department of Commerce, Document NSRDS-NBS 37, Washington, June 1971.

Suggested Reading

Ashton, J. D. and Holditch, J. E. R. (1975) 'Homogenized oil injection at DOFASCO', in *Ironmaking Proceedings,* Vol. 34, Toronto, 1975, AIME, New York, pp. 261–283.

Fletcher, L. N. and Garbee, A. K. (1975) 'Coal for blast furnace injection', in *Energy Use and Conservation in the Metals Industry,* Chang, Y. A., Danver, W. M. and Cigan, J. M., Editors, AIME, New York, pp. 203–217.

Lunn, H. G. and Waterhouse, G. W. (1976) 'Fuel-oil injection into blast furnaces, a literature review', *Journal of the Institute of Fuel,* XLIX(399), 70–78.

Standish, N. (Editor) (1972), *Blast Furnace Injection,* The Australasian Institute of Mining and Metallurgy, Wollongong, Australia.

Woolf, P. L. (1977) 'Blast furnace practice – I', in *Blast Furnace Ironmaking 1977,* Lu, W. K., Editor, McMaster University, Hamilton, Canada, pp. 16–1 to 16–18.

Chapter 11 Problems. *Injectant Heat Demands, Permissible Levels, Effect on Blast Furnace Parameters*

11.1 Calculate the injectant heat demands of the following materials, all except (f) and (g) being injected at 298 K. Heat demand calculations for complex materials are illustrated in Appendix II.

 (a) propane (assume 100% C_3H_8);
 (b) natural gas (90 vol.% CH_4, 5% C_2H_6, 1% CO_2, 4% N_2) (heat of combustion, – 39 300 kJ per Nm^3);
 (c) bunker C oil (86 wt.% C, 14% H) (heat of combustion, –46 600 kJ per kg);
 (d) bituminous coal (77 wt.% C, 5% H, 6% O, 1% S, 2% N, 9% SiO_2) (heat of combustion, –32 000 kJ per kg);
 (e) reformed natural gas (32 vol.% CO, 64% H_2, 1% CO_2, 3% N_2);
 (f) O_2 (1450 K), i.e. in 1450 K blast;
 (g) H_2O (1500 K).

The Effects of Tuyère Injectants 151

Indicate clearly by the symbols and units of your answers how the demand term should be incorporated into the blast-furnace-operating equations.

It is suggested that problems 11.2, 11.3 and 11.4 be solved by a computer programme put together from those developed in Chapters 3, 4, 7 and 9, appropriately modified. Solution by a numerical simultaneous equation technique might be the simplest method (e.g. to solve equations (11.1), (11.8) and (11.12)).

11.2 A Mexican blast furnace is operating with 1400 K dry air blast and 20 Nm³ of C_3H_8 (298 K) per tonne of product Fe. Previous experience with the furnace indicates that under its current operating conditions the wustite reduction zone heat <u>demand</u> is 400 000 kJ per kg mole of Fe (exclusive of injectant heat demands).

Calculate for these conditions, the minimum coke (89% C) requirement of the furnace per tonne of product Fe (assume that the metal product is 95% Fe, 5% C).

11.3 The wustite reduction zone heat demand (excluding the heat demands of H_2O in blast and other injectants) of the new blast furnace described in Problem 3.2 (b) is 385 000 kJ per kg mole of product Fe. Calculate, for the condition of maximum CH_4 injection,

(a) the coke (89% C) requirement per tonne of product Fe;
(b) the moist blast air requirement per tonne of product Fe.

The analysis of the hot metal is 95% Fe, 5% C (ignore other impurities).

11.4 The operators of the blast furnace described in Problem 9.1 are thinking of improving its operation by injecting oxygen and hydrocarbons through the tuyères. The plant has 25 Nm³ of pure oxygen available per tonne of Fe produced by the furnace.

Calculate:

(a) the maximum amount of bunker C oil (specified in Problem 11.1(c)) which could be injected into the furnace (i) per 1000 Nm³ of dry blast air and (ii) per tonne of Fe (experience having shown that the tuyère flame temperature of this furnace must be maintained above 2400 K for satisfactory operation);
(b) the effect that this maximum would have on the coke and blast requirements of the process;
(c) the increase in productivity which could be obtained with this maximum oil injection if the Problem 9.1 furnace is already operating at maximum blower capacity. Assume that the Fe production rate is proportional to the ratio:

$$\left\{ \frac{\text{blast rate (Nm}^3 \text{ of air + pure O}_2 \text{ per minute)}}{\text{blast requirement (Nm}^3 \text{ of air + pure O}_2 \text{ per tonne of Fe)}} \right\}.$$

11.5 The superintendent of stoves for the Problem 11.4 blast furnace wishes to know what top-gas fuel value he can expect under the proposed bunker C oil injection conditions. To predict this he needs to know the top-gas composition. Make this composition prediction for him. Assume that CO and H_2 have an equal opportunity to reduce Fe_2O_3/Fe_3O_4 to $Fe_{0.947}O$.

CHAPTER 12

Addition of Details into the Operating Equations: Heat Losses; Reduction of Si and Mn; Dissolution of Carbon; Formation of Slag; Decomposition of Carbonates

Chapter 11 showed how tuyère injectants are incorporated into the basic blast-furnace-operating equations. This incorporation of injectants was the final major step in developing the equations and only a few details are now needed to complete the model. These final details are:

(a) stoichiometric effects of SiO_2 and MnO reduction;
(b) stoichiometric effects of $CaCO_3$ and $MgCO_3$ decomposition;
(c) enthalpy effects of heat losses, SiO_2 and MnO reduction, carbon dissolution, slag formation and $CaCO_3$ decomposition.

12.1 Stoichiometric Effects

12.1.1 Reduction of SiO_2 and MnO

SiO_2 and MnO are partially reduced in the wustite reduction zone. This reduction results in a net transfer of oxygen from SiO_2 and MnO to the reducing gases and thus it has a significant effect on the wustite reduction zone oxygen balance. The usual levels of Si and Mn in the product iron are about 1 wt.% which is equivalent to:

$(Si/Fe)^m = 0.02$ moles of Si per mole of product Fe,
$(Mn/Fe)^m = 0.01$ moles of Mn per mole of product Fe.

Considering SiO_2 reduction first, it can be seen that each mole of reduced Si contributes 2 moles of O to the wustite reduction zone oxygen balance. Per mole of Fe, therefore, the oxygen contributed to the oxygen balance by Si reduction is

$$2(Si/Fe)^m.$$

Similarly, the oxygen contributed by MnO reduction is

$$1(Mn/Fe)^m.$$

(Manganese enters the furnace as Mn_3O_4, Mn_2O_3 or MnO_2 but it is in the form of MnO by the time it descends into the wustite reduction zone.)

These additional oxygen inputs alter the wustite reduction zone oxygen input equation (Sections 11.2, 11.3) to

$$n_O^{iwrz} = n_O^B + 1.06 + z \cdot n^I + 2(Si/Fe)^m + (Mn/Fe)^m$$

while the oxygen-output equation remains unaltered at:

$$n_O^{owrz} = 1.3 \cdot n_C^A + 0.38 \cdot y \cdot n^I.$$

Matching these input and output oxygen equations, we obtain the modified operating equation:

$$n_O^B + 1.06 + z \cdot n^I + 2(Si/Fe) + (Mn/Fe) = 1.3 \cdot n_C^A + 0.38 \cdot y \cdot n^I.$$

(12.1)

12.1.2 Stoichiometric Effects of Limestone Decomposition: (A) Effect on the Carbon Balance

Limestone decomposes according to the reaction:

$$CaCO_3 \rightarrow CaO + CO_2$$

at temperatures above 1250 K. This means that all limestone charged to the furnace descends into the wustite reduction zone where it splits up into CaO and CO_2. The CaO subsequently forms a slag with SiO_2 and other gangue components, and hence it leaves the furnace without affecting the stoichiometry of the reduction process.

The CO_2 component is, on the other hand, transferred from solid $CaCO_3$ to the gas phase and hence it affects the wustite reduction zone

Addition of Details into the Operating Equations

oxygen and carbon balances, i.e. it is 'active'. It behaves somewhat like a tuyère injectant.

Excluding its 'inert' CaO component, each mole of $CaCO_3$ charged to the furnace results in a net input of 1 mole of C and 2 moles of O into the wustite reduction zone. Per kg mole of product Fe, these inputs amount to

$$\text{carbon input} = n_{CaCO_3},$$
$$\text{oxygen input} = 2n_{CaCO_3},$$

where n_{CaCO_3} represents kg moles of $CaCO_3$ charged to the furnace per kg mole of product Fe.

In terms of the wustite reduction zone carbon balance, the addition of n_{CaCO_3} kg moles of carbon to the wustite reduction zone alters the carbon balance equation

$$n_C^{iwrz} = n_C^{owrz} \quad (4.2 \text{ wrz})$$

to

$$n_C^{coke} + x \cdot n^I + n_{CaCO_3} = n_C^A + (C/Fe)^m \quad \text{(Section 11.2)}$$

or

$$n_C^A = n_C^{coke} + x \cdot n^I + n_{CaCO_3} - (C/Fe)^m. \quad (12.2)$$

Equation (12.2) is the final operating equation for calculating the coke requirement of a furnace (n_C^{coke}) from its demand for active carbon (n_C^A) and known amounts of (i) carbon-in-iron and (ii) carbon from non-coke sources.

12.1.3 Limestone Decomposition: (B) Effect on the Oxygen Balance

In oxygen-balance terms, limestone decomposition alters the oxygen-input equation by $2n_{CaCO_3}$ moles of O (neglecting the 'inert' CaO component) but leaves the form of the oxygen output equation unchanged. The wustite reduction zone oxygen balance becomes, therefore,

$$n_O^B + 1.06 + z \cdot n^I + 2(Si/Fe)^m + (Mn/Fe)^m + 2n_{CaCO_3}$$
$$= 1.3 \cdot n_C^A + 0.38 \, y \cdot n^I \quad (12.3)$$

where n_C^A has the precise meaning given by equation (12.2).

12.1.4 Overall Effects of $CaCO_3$

Charging of limestone always leads to an increased carbon requirement in the wustite reduction zone and this must be met by increasing the coke and/or hydrocarbon supply to the furnace. As is shown later (Section 12.2), this is due to enthalpy as well as stoichiometric effects. In stoichiometric terms it can be visualized that limestone brings in a gas of higher O/C ratio (CO_2 : O/C = 2) than the gas which ultimately ascends from the wustite reduction zone, $(O/C)^{gwrz} = 1.3$. This means that a portion of the CO_2 from limestone oxidizes carbon in the wustite reduction zone, thereby increasing the coke and/or hydrocarbon demand of the process.

12.1.5 Other Carbonates: Dolomite and Magnesite

MgO tends (Elliott, 1977; Lowing, 1977) to lower the viscosity of blast-furnace slags and also the sensitivity of their viscosity to composition changes. For this reason it is usually included in blast-furnace charges either as MgO itself in self-fluxing sinter or steelmaking slag, or separately as dolomite.

Dolomite decomposes by the reaction

$$MgCO_3 \cdot CaCO_3 \rightarrow CaCO_3 + MgO + CO_2$$

at approximately 1000 K (Manning, 1969). Likewise magnesite decomposes to its oxide and CO_2, i.e.

$$MgCO_3 \rightarrow MgO + CO_2$$

at approximately 900 K.

These low decomposition temperatures indicate that $MgCO_3$ decomposes before it can descend into the wustite reduction zone and that, as a consequence, it has no effect* on (i) the wustite reduction zone operating equations or (ii) predictions of n_C^A, n_O^B or n^I. Unlike limestone, therefore $MgCO_3$ does not increase carbon consumption. Unfortunately MgO is not as good a chemical desulphurizer as CaO so that the amount of $MgCO_3$

---

*Dolomite and magnesite do, however, affect top-gas composition and temperature, and for this reason $MgCO_3$ decomposition must be represented in equations (4) and (10), the whole furnace stoichiometric and enthalpy equations.

Addition of Details into the Operating Equations

which can be charged to a furnace is limited, i.e. the MgO:CaO weight ratio in blast-furnace slags is usually less than 1:3.

12.2 Enthalpy Effects

Heat losses from the furnace, reduction of minor components, decomposition of $CaCO_3$, and formation of liquid slag all affect the wustite reduction zone enthalpy balance. The following subsections show how the effects of these items can be included in the operating equations as additional discrete heat demand terms D^{loss}, D^{Si}, D^{Mn}, D^{CaCO_3} and D^{slag}. The enthalpy effect of carbon dissolution in iron is also discussed.

12.2.1 Heat Losses from the Furnace by Radiation and Convection: D^{loss}

Heat-loss data from large modern industrial furnaces (Flierman, 1977) suggest that radiative and convective losses from the iron blast furnace are of the order of*

$$8 \times 10^6 \; (kJ \; m^{-1} \; hr^{-1}) \times \text{hearth diameter (m)} \qquad (12.4)$$

of which about 80% is lost through the bosh and lower shaft (wustite reduction zone). Thus radiative and convective heat losses from the wustite reduction zone, expressed in kJ per kg mole of product Fe, may be represented by

$$Q = D^{loss} = \frac{6.4 \times 10^6 \; kJ \; m^{-1} \; hr^{-1} \times \text{hearth diameter (m)}}{\left\{\dfrac{\text{pig iron production rate (kg hr}^{-1})}{55.85}\right\} \times \left\{\dfrac{\text{\% Fe in pig iron}}{100}\right\}}$$

(12.5)

Industrial heat losses per kg mole of product Fe as calculated from the data in Table 1.1 (assuming a 93% Fe pig iron) are shown in Table 12.1.

---

*Flierman's actual equation is:

$$\text{Furnace heat loss} = 5.4 \times 10^6 \times \text{hearth diameter} + 0.85 \times 10^6 \times \text{number of tuyères} \quad kJ \; hr^{-1}.$$

Equation (12.4) is interpreted from this equation by noting that modern blast furnaces have 2½ to 3 tuyères for each meter of hearth diameter (Nakamura, 1978).

TABLE 12.1

Hearth diameter (m)	Wustite reduction zone heat loss (eqn. (12.4)) (kJ hr^{-1})	Fe Production rate (tonnes hr^{-1})	Fe Production rate (kg moles hr^{-1})	Dloss = Q (eqn. (12.5)) (kJ (kg mole of Fe))$^{-1}$
8.5	54 × 10^6	100	1700	32 000
10.7	68 × 10^6	220	4000	17 000
11.1	71 × 10^6	190	3400	21 000
14.0	90 × 10^6	310	5600	16 000
14.3	92 × 10^6	380	6900	13 000

Addition of Details into the Operating Equations

These data show that the wustite reduction zone heat demand due to convective and radiative heat losses is of the order of

$$D^{loss} = Q = 20\,000 \text{ kJ (kg mole of product Fe)}^{-1}.$$

D^{loss} depends significantly on the productivity of each blast furnace (specifically on hearth diameter ÷ production rate) and hence it should be calculated for each specific blast-furnace operation.

12.2.2 Reduction of SiO_2: D^{Si}

Reduction of SiO_2 to Si and its dissolution in liquid iron exerts a considerable heat demand (D^{Si}) on the wustite reduction zone. It is made up of three components:

(a) *Reduction demand:*

$$SiO_2 \xrightarrow[(1200 \text{ K})]{} Si_s + O_2 \qquad \text{Demand} = -H^f_{1200} \text{ SiO}_2$$

$$= +901\,000 \text{ kJ (kg mole of Si)}^{-1}.$$

(b) *Heating and melting demand:*

$$Si_s \xrightarrow[1200]{} Si_l \text{ at } 1800 \qquad \text{Demand} = [H^\circ_{1800} - H^\circ_{1200}]_{Si_l \quad Si_s}$$

$$= +67\,000 \text{ kJ (kg mole of Si)}^{-1}.$$

(c) *Dissolution demand:*

$$Si_l \xrightarrow[1800]{} [Si]_{\text{in iron at }1800} \qquad \text{Demand} = H^M_{Si}$$

$$= -95\,000 \text{ (exothermic)}$$
$$\text{kJ (kg mole of Si)}^{-1}$$
$$(5\% \text{ C}, 1\% \text{ Si, Healy, 1977}).$$

Thus the total heat demand for Si is

$$D^{Si} = +873\,000 \text{ kJ (kg mole of dissolved Si)}^{-1}$$

and the additional demand term, kJ per kg mole of product Fe, in the enthalpy balance equations (8a), (11.12) is

$$(Si/Fe)^m \cdot D^{Si} = (Si/Fe)^m \cdot 873\ 000 \text{ kJ (kg mole of product Fe)}^{-1}$$
(12.6)

As indicated in Section 12.1, $(Si/Fe)^m$ is in the order of 0.02, so that the heat demand for silicon $(Si/Fe)^m \cdot D^{Si}$ is typically 17 000 kJ per kg mole of product Fe.

It will be noted that the oxygen resulting from SiO_2 decomposition is left at 1200 K. The usefulness of this approach was demonstrated by equations (8.4b) and (11.10) in which the $H°_{1200}$ terms are shown to always disappear by virtue of the stoichiometric oxygen balance. For this reason the same approach is used here for minor impurities.

12.2.3 Reduction of MnO: D^{Mn}

The three heat demands for MnO reduction and dissolution

$$MnO_{1200} \longrightarrow [Mn]_{\text{in iron} \atop 1800} + \tfrac{1}{2} O_2{}_{1200}$$

are

(a) reduction $+386\ 000$ kJ (kg mole of Mn)$^{-1}$,

(b) heating and melting $+45\ 000$ kJ (kg mole of Mn)$^{-1}$,

(c) dissolution of liquid Mn in Fe (Hultgren, 1973) $+4000$ kJ (kg mole of Mn)$^{-1}$,

for a total heat demand of

$$D^{Mn} = 435\ 000 \text{ kJ (kg mole of Mn)}^{-1}.$$

As shown above for silicon, the manganese heat demand term in the enthalpy balance operating equations is

$$(Mn/Fe)^m \cdot D^{Mn} = (Mn/Fe)^m \cdot 435\ 000 \text{ kJ (kg mole of dissolved Mn)}^{-1}$$
(12.7)

and it typically has a value $[(Mn/Fe)^m = 0.01]$ of 4000 kJ per kg mole of product Fe.

Addition of Details into the Operating Equations

12.2.4 Dissolved Carbon

The wustite reduction zone heat demand exerted by dissolved carbon has been used many times in previous chapters but its basis of calculation has not been discussed. It consists of two demand components:

(a) *Heating demand*

$$C_S \underset{1200}{} \longrightarrow C_S \underset{1800}{} \qquad \text{Demand} = [H^\circ_{1800} - H^\circ_{1200}]_C = +14\,000 \text{ kJ (kg mole of C)}^{-1}.$$

(b) *Dissolution demand*

$$C_S \underset{1800}{} \longrightarrow [C]_{\text{in iron}} \underset{1800}{} \qquad \text{Demand} = H^M_C = +30\,000 \text{ kJ (kg mole of C)}^{-1}.$$

(5% C, Healy, 1977)

for a total heat demand of

$$D^C = 44\,000 \text{ kJ (kg mole of carbon)}^{-1}$$

Per kg mole of Fe, the total dissolved carbon heat demand is given by:

$$(C/Fe)^m \cdot D^C = (C/Fe)^m \cdot 44\,000 \qquad (12.8)$$

and it has a typical value (5% C; $(C/Fe)^m = 0.25$) of 11 000 kJ per kg mole of product Fe.

In the development of the enthalpy equations in Chapters 8, 9 and 11, this carbon heat demand was lumped in the term D^{wrz}. Removal of $(C/Fe)^m \cdot D^C$ from D^{wrz} leaves the enthalpy demand

$$- 1.06 \underset{Fe_{0.947}O}{H^f_{1200}} + [H^\circ_{Fe(l)} - H^\circ_{Fe(s)}]_{\substack{1800 \quad 1200}}$$

which is represented from now on as D^{Fe}; 319 000 kJ (kg mole of Fe)$^{-1}$.

12.2.5 Slag Heat Demand: D^{slag}

With the exception of calcium and magnesium carbonates, slag components do not affect the stoichiometry of reduction. They do, however, affect the enthalpy balances of the process. Expressions for slag heat demand are developed in detail in Appendix III. It is shown there that the heat demand per kilogram of slag can be represented by

$$D^{slag} = H^f_{1200\ slag} + [H_{1800} - H_{1200}]_{slag}$$

where $H^f_{1200\ slag}$ is the heat of formation of the slag at 1200 K from its components at 1200 K, and $[H_{1800} - H_{1200}]_{slag}$ is the enthalpy required to (i) heat the slag to 1800 K and to (ii) melt it, all kJ per kg of slag.

Typical values of $H^f_{1200\ slag}$ and $[H_{1800} - H_{1200}]_{slag}$ are -750 and $+1000$ kJ per kg of slag, respectively (Appendix III), giving a representative D^{slag} value of $+250$ kJ per kg of slag ($\pm 20\%$).

Typical industrial slag falls are 300 kg per tonne of Fe (18 kg of slag per kg mole of Fe), Table 1.1, which is equivalent to a slag heat demand of

$$wt_{slag} \cdot D^{slag} = 18 \cdot 250 = 5000 \text{ kJ (kg mole of product Fe)}^{-1}. \quad (12.9)$$

12.2.6 Heat for Dissociating $CaCO_3$: D^{CaCO_3}

As was pointed out in Section 12.1, limestone decomposes at about 1250 K, i.e. in the wustite reduction zone. This decomposition places an enthalpy burden on the wustite reduction zone in excess of that which would be required if the CaO were added in the form of burnt lime (CaO).

The simplest way to treat limestone is to consider it as a 1200-K enthalpy input into the wustite reduction zone. Numerically, the input (per kg mole of product Fe) is

$$n_{CaCO_3} \cdot H^\circ_{1200\ CaCO_3}$$

which can be broken down into:

$$n_{CaCO_3} \cdot H^\circ_{1200\ CaCO_3} = n_{CaCO_3} \cdot (H^L_{1200\ CaCO_3} + H^\circ_{1200\ CaO} + H^\circ_{1200\ C} + H^\circ_{1200\ O_2})$$

$$(12.10)$$

Addition of Details into the Operating Equations

where $H^L_{CaCO_3}$ is the heat of combination for

$$CaO + C + O_2 \longrightarrow CaCO_3 \quad H^L_{1200} \atop CaCO_3 = -561\,000 \text{ kJ (kg mole of } CaCO_3)^{-1}.$$

Incorporation of the right side of equation (12.10) into enthalpy balance equation (11.10) leads, after rearranging, to*

$$D^{wrz} + n^I D^I + n_{CaCO_3} \cdot (-H^L_{1200} \atop CaCO_3)$$

$$= E^B \cdot n^B_O - n^A_C \cdot (0.7 \cdot H^f_{1200} \atop CO + 0.3 \cdot H^f_{1200} \atop CO_2) - 0.38 \cdot y \cdot n^I \cdot H^f_{1200} \atop H_2O \quad (12.11)$$

where n^A_C has the precise meaning of equation (12.2), i.e.

$$n^A_C = n^{coke}_C + x \cdot n^I + n_{CaCO_3} - (C/Fe)^m.$$

By reference to equation (AIII.1) it can be seen that the addition of $CaCO_3$ to the furnace *rather than CaO* leads to the extra term

$$n_{CaCO_3} \cdot (-H^L_{1200} \atop CaCO_3) \quad \text{kJ (kg mole of product Fe)}^{-1}.$$

on the demand side of the wustite reduction zone enthalpy-balance equation. In demand notation this term can be represented by

$$n_{CaCO_3} \cdot D^{CaCO_3}$$

or

$$D^{CaCO_3} = (-H^L_{1200} \atop CaCO_3) = +561\,000 \text{ kJ (kg mole of } CaCO_3)^{-1}. \quad (12.12)$$

This limestone demand has the precise meaning of being the extra heat demand required when CaO (for slagmaking) is supplied to the wustite reduction zone in the form of $CaCO_3$ rather than CaO.

A modern-day pellet- or ore-fed blast furnace typically uses 100 kg of $CaCO_3$ per tonne of product Fe. This is equivalent to approximately 0.05 kg moles of $CaCO_3$ per kg mole of product Fe (i.e. $n_{CaCO_3} = 0.05$) and

---

*Where D^{wrz} includes the slag heat demand.

thus to a heat demand of 30 000 kJ per kg mole of product Fe. The $CaCO_3$ load in furnaces which are charged with prefluxed sinter is, on the other hand, very small because most of the lime for slagmaking is already present in the sinter as CaO. Sinter-charging avoids, therefore, most of the $CaCO_3$ heat demand.

12.3 Summary

The stoichiometric effects of (i) reduction and dissolution of minor components in the product iron and (ii) decomposition of carbonates have been incorporated into the wustite reduction-zone-operating equations. Si and Mn reduction and $CaCO_3$ decomposition have been shown to significantly alter the operating equations. Magnesite and the $MgCO_3$ part of dolomite decompose high in the stack (i.e. at temperatures <1000 K), however, and hence they do not affect the wustite reduction-zone-operating equations or the furnace-operating parameters.

The enthalpy effects of: heat losses; reduction of SiO_2 and MnO; formation of slag and decomposition of limestone are clearly identifiable in the form of a series of heat demand terms. All of these processes exert enthalpy demands on the wustite reduction zone and hence they require carbon and blast in addition to that required for iron reduction only. Wustite reduction zone heat demands are summarized by the equation:

$$D^{wrz} = D^{Fe} + (C/Fe)^m \cdot D^C + D^{loss} + (Si/Fe)^m \cdot D^{Si} + (Mn/Fe)^m \cdot D^{Mn}$$
$$+ wt._{slag} \cdot D^{slag} + n_{CaCO_3} \cdot D^{CaCO_3} \qquad (12.13)$$

excluding tuyère injectant-heat demands. Typical total wustite reduction zone heat demands, D^{wrz}, of large modern blast furnaces are in the range of 360 000 to 400 000 kJ per kg mole of product Fe.

References

Elliott, J. F. (1977) 'Blast furnace theory', in *Blast Furnace Ironmaking 1977*, Lu, W. K., Editor, McMaster University, Hamilton, Canada, pp. 2–8, 2–9.

Flierman, G. A. and Homminga, F. (1977) 'A comparison of BF-operating results obtained with sinter or pellet burden', in *Agglomeration 77*, Sastry, K. V. S., Editor, AIME, New York, p. 829.

Addition of Details into the Operating Equations

Healy, G. W. and McBride, D. L. (1977) 'The [BOF] mass-energy balance', in *BOF Steelmaking*, Pehlke, R. D., Porter, W. F., Urban, R. F. and Gaines, J. M., Editors, AIME, New York, Chapter 13, p. 129.

Hultgren, R., Desai, P. D., Hawkins, D. T., Gleiser, M. and Kelley, K. K. (1973) *Selected Values of the Thermodynamic Properties of Binary Alloys*, American Society for Metals, Metals Park, Ohio, p. 840.

Lowing, J. (1977) 'The diagnostic approach to overcoming blast furnace operational problems', in *Ironmaking Proceedings*, Vol. 36, Pittsburgh, 1977, AIME, New York, p. 230.

Manning, F. S. and Philbrook, W. O. (1969) 'Rate phenomena', in *Blast Furnace-Theory and Practice*, Strassburger, J. H., Editor, Gordon & Breach, New York, pp. 878–881.

Nakamura, N., Togino, Y. and Tateoka, M. (1978) 'Behaviour of coke in large blast furnaces', *Ironmaking and Steelmaking*, 5(1), 7.

Chapter 12 Problems. *Heat Demands and Stoichiometric Effects of Minor Reactions*

Note: These problems are most easily solved by developing a series of readily identifiable linear equations (including equations for calculating slag weight and D^{wrz} and for representing injectant quantities per Nm^3 of blast) followed by a computer solution of the equations.

12.1 A small (9 m hearth diameter) blast furnace is operating under the following conditions:

Item	Value
(a) Metal production rate	2000 tonnes day^{-1}
(b) Temperature of blast air	1300 K
(c) Metal composition	5% C, 1.5% Si, 1% Mn
(d) Metal temperature	1800 K
(e) Ore composition	wt.Fe_2O_3/wt.SiO_2 = 19/1; plus manganese oxides
(f) Coke composition	88% C, 12% SiO_2
(g) Flux	$CaCO_3$
(h) Slag composition and heat demand	wt. CaO/wt. SiO_2 = 1.18, D^{slag} = 250 kJ per kg. (All Mn in the ore is reduced to metal)
(i) Tuyère injectants	blast moisture only (8 g per Nm^3 of moist blast)

Predict, based on these data, the minimum coke demand (per tonne of Fe) which would be expected for this operation. Determine also the amounts of flux and moist blast which would be required.

12.2 One advantage of feeding blast furnaces with sinter rather than pellets is that the limestone flux can be calcined to CaO during the sintering process, i.e. before feeding the ore to the blast furnace. A fluxed sinter feed eliminates, therefore, the enthalpy and stoichiometric demands exerted by $CaCO_3$ decomposition in the wustite reduction zone of the furnace. Calculate the changes in coke and blast demands of the Problem 12.1 blast furnace if the feed were to be changed from pellets to sinter (same SiO_2 content per tonne of Fe).

12.3 Sulphur is sometimes removed from molten pig iron (by magnesium or calcium carbide treatment) after the iron has been tapped from the blast furnace. This procedure lowers the CaO requirement of the slag and hence it lowers the amount of limestone flux which must be added to the furnace. To show the effect of external desulphurization, recalculate Problem 12.1 with the CaO/SiO_2 requirement lowered to 0.9. How much would the production rate be increased by this change, assuming that production rate is inversely proportional to the volume of moist blast air required per tonne of Fe? Assume that D^{slag} remains at 250 kJ per kg of slag.

12.4 In an experimental programme, prereduced iron pellets are included in the Problem 12.1 charge. In a particular test, 20% of the product Fe originates from these pellets. Predict the improvements in coke demand and production rate which can be expected from this test. Assume that all the iron in the pellets is metallic and that the amounts of SiO_2 and MnO brought into the furnace (per unit of Fe) are the same for the prereduced pellets as for the iron ore.

12.5 Phosphorus is an important impurity in many European iron ores. The metal product from Belgian blast furnaces can contain, for example, up to 1½% P. Ascertain the heat demand of dissolved phosphorus and show how this impurity can be represented in the operating equations. Phosphorus can be considered to enter the wustite reduction zone as $(CaO)_3 \cdot P_2O_5$. Its enthalpy of mixing in Fe (1800 K) can (Healy, 1977) be represented by

$$\tfrac{1}{2}P_2 \underset{180}{} \longrightarrow \underset{1800}{P_{\text{in Fe}_l}} \qquad H_P^M = -122\,000 \text{ kJ (kg mole of P)}^{-1}.$$

Ignore heating and melting of CaO, i.e. assume that these effects are taken into consideration in a slag-demand calculation.

CHAPTER 13

Summary of Blast-Furnace-Operating Equations: Comparison between Predictions and Practice

The previous nine chapters have shown in detail the concepts and methods involved in developing our predictive equations. This brief chapter

(a) summarizes the steps of the development and reproduces the most useful equations and graphical coordinates from each step;
(b) discusses a strategy by which the equations are adapted for computer calculation;
(c) compares predictions of the equations with the operating parameters of modern industrial blast furnaces.

It may be useful to recall at this point that the operating equations have all been developed for the bottom segment of the furnace, i.e. for the wustite reduction zone. However, continuity conditions have been adhered to throughout the development so that operating parameters calculated by the bottom segment equations must also be those for the whole furnace. The validity of the assumptions made during the development has been discussed in Chapter 10.

13.1 Summary of Model Development Steps

This section presents the stoichiometric, enthalpy and combined stoichiometric/enthalpy equations in (i) their most basic form, carbon and iron oxide charge only; (ii) their basic form with a tuyère injectant of the type $C_x(H_2)_y O_z$; (iii) a final form which includes the effects of slag, limestone, minor element reduction and heat losses. Graphical coordinates are also presented for each of these three cases.

The Iron Blast Furnace

	Equation number	Page

13.1.1 Oxygen Balance Stoichiometric Equations

Basic: $\quad n_O^B + 1.06 \quad = 1.3 \cdot n_C^A$ — 7 — 81

Injectant:* $n_O^B + 1.06 + z \cdot n^I = 1.3 \cdot n_C^A + 0.38 \cdot y \cdot n^I$ — 11.8 — 129

Final: $\quad n_O^B + 1.06 + z \cdot n^I + 2(Si/Fe)^m + (Mn/Fe)^m + 2n_{CaCO_3}$
$\quad = 1.3 \cdot n_C^A + 0.38 \cdot y \cdot n^I$ — 12.3 — 155

13.1.2 Carbon Balance Stoichiometric Equations

Basic: $\quad n_C^A + (C/Fe)^m = n_C^{coke}$ — 11.1 — 126

Injectant:* $n_C^A + (C/Fe)^m = n_C^{coke} + x \cdot n^I$

Final: $\quad n_C^A + (C/Fe)^m = n_C^{coke} + x \cdot n^I + n_{CaCO_3}$ — 12.2 — 155

13.1.3 Enthalpy Balance Equations (all heat demands are for the wustite reduction zone, Chapters 8 and 12)

Basic:† $\quad D^{Fe} + (C/Fe)^m \cdot D^C \quad = 198\,000 \cdot n_C^A + E^B \cdot n_O^B$ — 8 — 92

Injectant:* $D^{Fe} + (C/Fe)^m \cdot D^C + n^I \cdot D^I = 198\,000 \cdot n_C^A + E^B \cdot n_O^B + 95\,000 \cdot y \cdot n^I$ — 11.12 — 133

Summary of Blast-Furnace-Operating Equations

Final: $D^{Fe} + (C/Fe)^m \cdot D^C + n^I \cdot D^I + D^{loss} + (Si/Fe)^m \cdot D^{Si}$
$+ (Mn/Fe)^m \cdot D^{Mn} + wt_{slag} \cdot D^{slag}$
$+ n_{CaCO_3} \cdot D^{CaCO_3} = 198\,000 \cdot n_C^A + E^B \cdot n_O^B + 95\,000 \cdot y \cdot n^I$ from 11.12

or $D^{wrz} + n^I \cdot D^I = 198\,000 \cdot n_C^A + E^B \cdot n_O^B + 95\,000 \cdot y \cdot n^I$ 133

13.1.4 Combined Stoichiometric and Enthalpy Balance Equations (all heat demands are for the wustite reduction zone, Chapters 8 and 12)

Basic: $D^{Fe} + (C/Fe)^m \cdot D^C = n_C^A \cdot (198\,000 + 1.3E^B) - E^B \cdot \{1.06\}$ 92

Injectant:* $D^{Fe} + (C/Fe)^m \cdot D^C + n^I \cdot D^I = n_C^A \cdot (198\,000 + 1.3E^B) - E^B \cdot \{1.06 + z \cdot n^I\}$ 9
$+ 0.38 \cdot y \cdot n^I \cdot (249\,000 + E^B)$ from 11.15

Final: $D^{wrz} + n^I \cdot D^I = n_C^A \cdot (198\,000 + 1.3E^B) - E^B \cdot \{\}$ 146
$+ 0.38 \cdot y \cdot n^I \cdot (249\,000 + E^B)$

where $\{\} = \{1.06 + z \cdot n^I + 2(Si/Fe)^m + (Mn/Fe)^m$ 146, 155, 164
$+ 2n_{CaCO_3}\}$ from equations (12.2), (12.13), (11.15)

*Injectant formula, $C_x(H_2)_yO_z$. Complex injectants can be represented as shown in Appendix II.

†$E^B = \frac{1}{2} \cdot [H_{T_B}^\circ - H_{1200}^\circ]O_2 + \frac{1}{2} \cdot \frac{0.79}{0.21} \cdot [H_{T_B}^\circ - H_{1200}^\circ]N_2$

= enthalpy in blast air above 1200 K.

13.1.5 Wustite Reduction Zone Heat Demands (assuming a 1200 K thermal reserve)

Iron, 1800 K (from wustite): $D^{Fe} = 319\,000$ kJ (kg mole of Fe)$^{-1}$

Carbon (in iron): $D^C = 44\,000$ kJ (kg mole of C)$^{-1}$

Injectant (298 K):
$$D^I = -H_{298}^f \underset{C_x(H_2)_y O_z}{} + x \cdot [H_{1200}^o - H_{298}^o]_C$$
$$+ y \cdot [H_{1200}^o - H_{298}^o]_{H_2} + \tfrac{1}{2} \cdot z \cdot [H_{1200}^o - H_{298}^o]_{O_2}$$

kJ (kg mole of injectant)$^{-1}$

Final:
$$D^{loss} = \left\{ \frac{\text{rate of hearth and bosh heat loss (kJ hour}^{-1}\text{)}}{\text{rate of Fe production (kg moles hour}^{-1}\text{)}} \right\}$$

$\simeq 20\,000$ kJ (kg mole of Fe)$^{-1}$

$D^{Si} = 873\,000$ kJ (kg mole of Si)$^{-1}$

$D^{Mn} = 435\,000$ kJ (kg mole of Mn)$^{-1}$

$D^{slag} \simeq 250$ [kJ (kg of slag)$^{-1}$]

$D^{CaCO_3} = 561\,000$ kJ (kg mole of CaCO$_3$)$^{-1}$

Equation number	Page
12.8	161
12.5	157
12.6	160
12.7	160
12.9	162
12.12	163

Equilibrium Point W, Enthalpy Point H, and the O/C = 0 Intercept; Meanings of the Slope of the Operating Line

Equilibrium point W

The coordinates of equilibrium point W are:

$$O/C = 1.3 \quad : \quad O/Fe = 1.06$$

These coordinates assume that $Fe_{0.947}O/Fe/CO/CO_2$ equilibrium is achieved in a 1200-K thermal reserve zone. The effects of higher or lower thermal reserve temperatures are discussed in Section 10.7 and the effects of non-attainment of equilibrium are discussed in Section 10.6 (Fig. 10.1).

Other important points and meanings

Point H

	O/C	O/Fe
Basic	$1.06 - \dfrac{169\,000}{282\,000 + E^B}$	$\dfrac{D^{Fe} + (C/Fe)^m \cdot D^C + E^B \cdot \{1.06\}}{282\,000 + E^B}$
Injectant	$1.06 - \dfrac{169\,000}{282\,000 + E^B}$	$\dfrac{D^{Fe} + (C/Fe)^m \cdot D^C + E^B \cdot \{1.06 + z \cdot n^I\} + n^I \cdot D^I - 0.38 \cdot y \cdot n^I \cdot (249\,000 + E^B)}{282\,000 + E^B}$
Final	$1.06 - \dfrac{169\,000}{282\,000 + E^B}$	$\dfrac{D^{wrz} + E^B \cdot \{\quad\} + n^I \cdot D^I - 0.38 \cdot y \cdot n^I \cdot (249\,000 + E^B)}{282\,000 + E^B}$

where: $\{\quad\} = \{1.06 + z \cdot n^I + 2(Si/Fe)^m + (Mn/Fe)^m + 2n_{CaCO_3}\}$

171

	Slope of operating line	O/Fe intercept at O/C = 0	Graph number	Page
Basic	$n_C^A = n_C^{coke} - (C/Fe)^m$	$-(n_O^B)$	9.1, 9.2	97, 99
Injectant	$n_C^A = n_C^{coke} - (C/Fe)^m + x \cdot n^I$	$-(n_O^B + z \cdot n^I - 0.38 \cdot y \cdot n^I)$	11.3, 11.4	139, 147
Final	$n_C^A = n_C^{coke} - (C/Fe)^m + x \cdot n^I + n_{CaCO_3}$	$-\left\{ \begin{array}{l} n_O^B + z \cdot n^I - 0.38 \cdot y \cdot n^I \\ + 2(Si/Fe)^m + (Mn/Fe)^m + 2n_{CaCO_3} \end{array} \right\}$	Chapter 12	

Summary of Blast-Furnace-Operating Equations

13.2 A Strategy for Computer Calculation

The operating equations presented above are readily amenable to computer calculation which

(a) permits the complex features of actual blast-furnace operations to be incorporated into the model;
(b) facilitates comparisons between prediction and practice.

An important potential use of the blast-furnace model is process optimization by linear programming (Chapter 14) and this is also facilitated by arranging the equations in computer-programme form.

13.2.1 A Computer Programme

The flow diagram of a computer programme for comparing prediction with practice is shown in Fig. 13.1. In basic terms, the programme reads burden and tuyère data from which it predicts the quantities of coke, total fuel and blast which are required for steady-state operation of the furnace. Additional input data are metal composition, slag basicity and estimates of convective and radiative heat losses. Tuyère injectants are usually specified per Nm^3 of blast but they may be expressed per tonne of hot metal for prediction/practice comparison purposes. Flame temperature, top-gas composition and top-gas temperature are also calculated.

13.3 Comparison of Model Predictions with Industrial Blast-furnace Data

Comparisons between the predictions of the Fig. 13.1 programme and published operating data from five modern blast-furnace operations are presented in Table 13.1. Inspection of the table shows that the predicted amounts of coke, total fuel and blast are well within 5% of their industrial counterparts.

The only parameter for which prediction differs significantly from practice is top-gas temperature. This discrepancy may be due to an erroneous basis of comparison, i.e. the actual temperatures may be measured near the sidewalls while the predictions are for average gas temperatures at the stockline. This situation would lead, as noted in

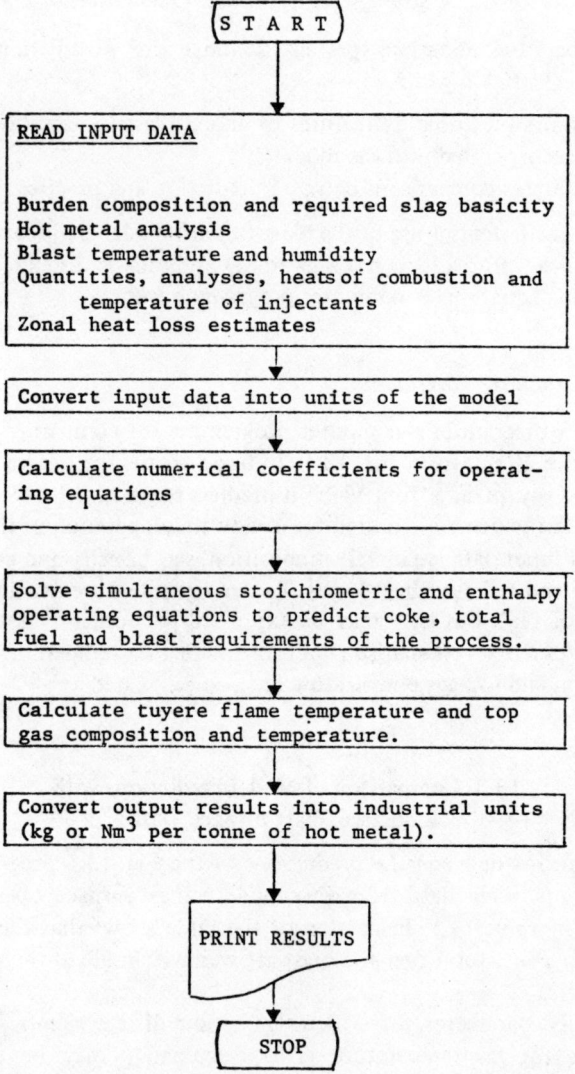

Fig. 13.1. Flow diagram of programme for testing predictions against practice. The simultaneous equations are solved by the Crout reduction direct method (Perry, 1973). Time of calculation: 0.1 second.

Comparisons between Actual and Predicted Coke, Total Fuel and Blast Requirements for Five Modern Blast Furnaces
Flame temperature and top-gas composition and temperature are also compared. The data for Fukuyama number 5 and Italsider number 5 are from Higuchi (1977) and De Marchi (1978). The remaining data were supplied directly by the blast-furnace operators.

	Furnace		Burns Harbour 'D'		Canadian		Fukuyama number 5		Chiba number 5		Italsider number 5	
	Data period		July 1977		1977		1976		Oct. 1976		1976	
FURNACE DATA	Metal production	tonnes per day	5700		2500		9900		4900		8100	
	Hearth diameter	m	10.7		8.5		14.4		11.1		14.0	
INPUT DATA	Blast temperature	K	1360		1340		1550		1330		1520	
	Blast moisture	g H_2O per Nm^3 blast	30		18		11		12		15	
	Oxygen enrichment	Nm^3 per tonne H.M.	20		0		10		30		15	
	Oil supply	kg per tonne H.M.	75		120*		55		40		60	
	Slag CaO/SiO_2 weight ratio		1.1		1.0		1.2		1.2			
			Actual	Predicted	Actual	Predicted	Actual	Predicted	Actual	Predicted	Actual	Predicted
COMPARATIVE DATA	Dry coke supply	kg per tonne H.M.	415	415	440	430	405	395	430	420	420	405
	Total fuel supply	kg per tonne H.M.	490	490	560	550	460	450	470	460	475	460
	Blast supply	Nm^3 per tonne H.M.	1220	1190	1690	1630	1040	1000	1040	1040	1020	1000
	Slag weight	kg per tonne H.M.	290	270	210	200	320	320	300	310		380
	Adiabatic flame temperature†	K	2260	2240	2170	2130	2460	2460	2470	2400		2420
	Top-gas temperature	K	420	580	470	680	400	470	400	410		390
	Top-gas analysis (dry basis) CO%		20	21	22	20	22	21	23	22	21	22
	CO_2%		20	22	16	18	22	22	22	23		22
	H_2%		5	4	5	4	3	3	3	2	3	3

*Includes 5 wt% H_2O.
† As defined in Appendix I.

Table 13.1, to predicted top-gas temperatures somewhat greater than those recorded in practice. However, with this exception, the agreement between prediction and practice is well within the expected accuracy of the model (±5%, Section 10.7).

13.4 Effects of Blast Temperature, Tuyère Injectants, Metallized Ore and Metal Impurities on Coke and Blast Requirements: Prediction and Practice

The Fig. 13.1 computer programme has also been used to predict the effects which various tuyère inputs will have on the coke and blast requirements of a blast furnace. The predictions and their published industrial counterparts are presented in Table 13.2.

It can be seen that predictions as to the effects of blast temperature and tuyère injectants on coke demand are all within the range of published industrial values. This may be one of the most important strengths of the model, because it permits the equations to be used with confidence for blast-furnace optimization.

The industrial and predicted data in Table 13.2 confirm that:

(a) coke demand per tonne of Fe is lowered by hydrocarbon injection and increased blast temperature;
(b) blast requirement is lowered by oxygen injection and increased blast temperature.

As the table also shows, hydrocarbon injection always requires extra blast (or oxygen) per tonne of Fe (Section 11.7) and oxygen injection usually requires extra carbon per tonne of Fe (Section 11.6), so that both have slight adverse effects on the process. However, oxygen and hydrocarbons are usually injected together so that these extra requirements are usually offsetting. The combined effects are readily predicted by the computer programme.

Prereduced, metallized ore is occasionally included in blast furnace burdens (Nakamura, 1974). The prereduced material is produced by direct reduction processes such as the Midrex or Stelco-Lurgi processes, and it contains mainly solid metallic Fe plus the gangue impurities of the original ore. The Fe does not need further reduction, and hence a partial charging of metallized ore to the blast furnace lowers the heat and coke demands of

Table caption (top):

Summary of... Effects of (i) Tuyere Injectants, (ii) Metallized Ore and (iii) Blast Temperature on the Coke and Blast Requirements of Large, Efficient Blast Furnaces

The actual data and industrial range information are from Nakamura (1974); the predicted data are based on the Fukuyama number 5 furnace described in Table 13.1.

Input variable	Unit of variable	Effect Coke requirement[1] kg (tonne H.M.)$^{-1}$ Predicted	Actual	Blast requirement N m^3 (tonne H.M.)$^{-1}$ Predicted	Actual	Industrial range of addition
Oil[2]	1 kg (tonne H.M.)$^{-1}$	−1.1	−1.0 to −1.4	+0.7		0 to 150 kg (tonne H.M.)$^{-1}$
Natural gas[3]	1 Nm^3 (tonne H.M.)$^{-1}$	−0.7	−0.5 to −2.0	+1.1		0 to 50 Nm^3 (tonne H.M.)$^{-1}$
Coal[4]	1 kg (tonne H.M.)$^{-1}$	−0.9	−0.8 to −1.1	+0.4		0 to 150 kg (tonne H.M.)$^{-1}$
Reformed gas[5]	1 Nm^3 (tonne H.M.)$^{-1}$	−0.25	−0.23 to −0.26	−0.4		0 to 300 Nm^3 (tonne H.M.)$^{-1}$
Oxygen[6]	1% enrichment	+1.0 to +1.2	'small penalty' (Woolf, 1977)	−42 to −50	−40	0 to 6%
Metallized ore[7]	10 kg (tonne H.M.)$^{-1}$	−3.4	−2 to −3	−7.5		0 to 300 kg (tonne H.M.)$^{-1}$
Blast temperature	100 K increase	−8 to −12	−8 to −20	−40 to −60		1200 to 1800 K

1. Dry coke composition: 87% C, 13% ash.
2. Oil composition: 86% C, 12% H; injection temperature = 298 K.
3. Natural gas composition: 96 vol.% CH_4; 1% CO_2; 3% N_2; injection temperature = 298 K.
4. Coal composition: 77% C, 5% H, 2% N, 5% O, 11% ash; injection temperature = 298 K.
5. Reformed gas composition: 72 vol.% H_2, 17% CO, 6% N_2, 1% CO_2, 2% H_2O, 2% CH_4; injection temperature = 1200 K.
6. Oxygen composition: 100 vol.% O_2; injection temperature = 1550 K.
7. Metallized ore composition: 95% Fe; 5% SiO_2.

the process. It can be noted, however, that the lowering of coke requirement attained in actual practice is somewhat smaller (Table 13.2) than that predicted by the model. This effect is probably due to some oxidation of the metallized ore during transport or even perhaps in the furnace itself.

13.4.1 Silicon, Manganese and Sulphur

The silicon and manganese dissolved in blast-furnace iron require (i) carbon for reduction and (ii) heat for reduction and melting. Changes in % Si and % Mn can be expected to have, therefore, significant effects on the carbon and blast requirements of the process. The Fig. 13.1 computer programme predicts the following quantitative effects:

Change	Change in fuel requirement	Change in blast-air requirement
− 0.1% Si	− 0.4%	− 0.5%
− 0.1% Mn	− 0.1%	− 0.15%

which show clearly that a deliberate lowering of % Si and % Mn will save fuel and will offer the possibility of an increased rate of iron production. Recent industrial practice has tended, in fact, towards lower levels of silicon and manganese in blast-furnace iron to take advantage of these effects. Low manganese levels are achieved principally by avoiding manganese-bearing ores but silicon is a more complex problem because all ores and cokes contain significant quantities of SiO_2.

The following factors are known to lower the Si content of blast-furnace metal:

(a) decreased blast temperature;
(b) decreased coke to ore ratio;
(c) decreased fuel to ore ratio;
(d) increased blast humidity;
(e) decreased slag acidity (i.e. lowered SiO_2 activity in slag).

A common feature of factors (a), (b), (c) and (d) in this list is that each of them either decreases the enthalpy supply to the furnace per tonne of Fe or increases the enthalpy demand ($D^I_{H_2O}$, case (d)). Thus each deprives

Summary of Blast-Furnace-Operating Equations

the furnace of enthalpy which it might otherwise use for reduction of SiO_2. In addition, factors (b) and (c) decrease the amount of reductant available for SiO_2 reduction.

Factors (a), (b), (c) and (d) also tend to cool the furnace, and this is probably the direct cause of their effect on silicon level. Lower furnace temperatures tend to decrease (i) the rate of Si reduction and (ii) the equilibrium % Si_{metal}/% $SiO_{2\,slag}$ ratio (Tsuchiya, 1976).

Generally speaking, blast furnaces are controlled so as to produce a steady supply of constant composition metal for subsequent refining to steel. A constant silicon content is especially important because silicon provides much of the 'fuel' for the steelmaking process (Ward, 1975). Silicon level in blast-furnace iron is most rapidly brought into line by adjusting blast temperature and humidity,* both of which produce almost immediately the effects described above (Walker, 1977). Coke to ore ratio, fuel to ore ratio and slag composition, which take effect much more slowly, can then be altered to make the silicon adjustment permanent.

Sulphur level in blast-furnace iron is controlled mainly by slag composition, rather than by thermal factors, so that it can be adjusted after conditions for the prescribed silicon content have been applied.

13.5 Summary

Reproduction of the predictive blast-furnace equations in condensed form has shown the effectiveness with which a gradual, structured build-up of complexity leads to a clear understanding of the blast-furnace process. This understanding is also facilitated by the demand/supply form of the enthalpy equations because this form permits the heat demands of each particular feature of the operation to be easily incorporated into the model. The equations are readily put into computer programme form which greatly simplifies comparisons between their predictions and actual industrial practice.

---

*Injection of hydrocarbons into the furnace without any other changes has the immediate effect of lowering flame temperature which causes metal temperature and % Si to fall. The lasting effect, however, is the opposite because the additional hydrocarbons supply additional enthalpy to the furnace, thereby eventually raising metal temperature and % Si. These two opposing factors make oil injection a rather poor parameter for controlling Si.

Coke, total fuel and blast inputs predicted by the model equations have been found to be in close agreement (±5%) with the actual inputs of large modern blast furnaces. The equations also accurately predict the changes in coke and blast requirements which result from hydrocarbon and oxygen injection, alteration of blast temperature and the use of metallized ore. These agreements between prediction and practice permit the model to be used with confidence for optimization purposes.

References

De Marchi, G., Fontana, P. and Tanzi, G. (1978) 'Blast furnace mathematical models, based on process theory, are suitable for high accuracy design, planning of operation policy and process control requirements', paper presented at 107th Annual AIME Meeting, Denver, Feb. 27, 1978.

Higuchi, M., Izuka, M., Kuroda, K., Nakayama, N. and Saito, H. (1977) 'Coke quality required for operation of a large blast furnace', Metals Society Conference, Middlesbrough, England, June 14–17, 1977, summarized in *Ironmaking and Steelmaking*, 4(5), 3–4.

Nakamura, N., Ishikawa, Y. and Tateoka, M. (1974) 'Measures to increase the productivity of blast furnace processes and plants and to reduce coke consumption in blast furnace', International Iron and Steel Congress, Dusseldorf, paper number 2.1.1.

Perry, R. H. and Chilton, C. H. (1973) *Chemical Engineers' Handbook*, McGraw Hill, New York, pp. 2–51.

Tsuchiya, N., Tokuda, M. and Ohtani, M. (1976), 'The transfer of silicon from the gas phase to molten iron in the blast furnace', *Metallurgical Transactions* B, 7B, 315–320.

Walker, H. (1977) 'Blast furnace practice-II', in *Blast Furnace Ironmaking, 1977*, Lu, W. K., Editor, McMaster University, Hamilton, Canada, pp. 17–29 to 17–33.

Ward, M. D. (1975) 'Consistent iron: the steelmaker's viewpoint', *Ironmaking and Steelmaking*, 2(2), 89–91.

Woolf, P. L. (1977) 'Blast furnace practice – I', in *Blast Furnace Ironmaking, 1977*, Lu, W. K., Editor, McMaster University, Hamilton, Canada, pp. 16–19.

CHAPTER 14

Blast-Furnace Optimization by Linear Programming

Blast-furnace operators are confronted with the on-going problem of deciding how, under varying economic and raw material supply conditions, to operate their furnaces so as to achieve a given objective. They may, for example, be asked at different times and under differing economic conditions to:

(a) minimize hot metal cost;
(b) maximize profit;
(c) minimize coke consumption;
(d) maximize productivity;

with any given combination of available raw materials and equipment. It is this general type of problem which can be attacked by linear programming with the aid of the operating equations of this text.

Linear programming can also be a useful tool for choosing the optimum size of plant (number and size of furnaces) and type of equipment (stove capacity and injection apparatus) for the attainment of a given objective. For example, it can be used to decide whether it is more economic to increase hot metal capacity by adding to existing facilities (e.g. by adding oxygen enrichment and oil-injection equipment to existing furnaces) or whether a new blast furnace should be built. Of course these decisions carry with them enormous economic implications.

For linear programming to be successful:

(a) the item to be optimized must be related to the operating variables in a linear manner;
(b) the operating variables must bear linear relationships to each other.

Fortunately, these linear requirements are met by the operating equations of this text and by blast-furnace cost and physical limitation relationships, so that linear programming is readily applied to blast-furnace problems.

14.1 A Simplified Optimization Problem

A commonplace purpose of blast-furnace optimization programmes is to determine which combination of raw materials, acceptable to the process, will result in a minimum cost for a specific aspect of the process. The objective of the programme might be, for example, to ascertain which fuel combination (coke plus tuyère-injected hydrocarbons) will result in a minimum total fuel cost per tonne of metal. This type of programme is relatively straightforward and hence it is useful for demonstration purposes as the following problem shows.

For the demonstration, consider the case of a blast furnace which can be supplied with two types of coke;

(a) lump coke (assume pure carbon) which is added with the ore charge and which costs $100 per tonne of carbon ($1.20 per kg mole of carbon);
(b) coke fines (assume pure carbon) which are injected through the tuyères and which cost $25 per tonne of carbon ($0.30 per kg mole of carbon).

Both types of fuel are readily available.

The problem is to determine the combination of these two fuels which will give, consistent with the operating requirements of the process, a minimum fuel cost per tonne of product Fe. The operating data for this particular furnace are shown in Table 14.1.

The quantities of lump coke and tuyère-injected coke which will maintain the process in stoichiometric and thermal balance are most clearly indicated by the equations:

$$n_C^A + (C/Fe)^m = n_C^{coke} + x \cdot n^I \qquad (11.1)$$

and

$$D^{wrz} + n^I \cdot D^I = S^{wrz}$$
$$= 198\,000 \cdot n_C^A + E^B \cdot n_O^B + 95\,000 \cdot y \cdot n^I. \qquad (11.12)$$

TABLE 14.1

Item	Specification	Quantity		Model variable
		kg per tonne of Fe	kg moles per tonne of Fe	
Fe		1000	17.9	
Iron oxide entering wustite reduction zone	$Fe_{0.947}O$			$(O/Fe)^{xwrz} = 1.06$
Pig iron	5% C	53 (carbon)	4.4	$(C/Fe)^m = 0.25$
Blast (dry air) temperature				$T_B = 1200$ K
Blast enthalpy				$E^B = 0$
Injectant	Coke fines 298 K (pure carbon)			$n^I = ?$ $x = 1$ $y = 0$ $z = 0$
Injectant heat demand				$D^I = 16\,000$ kJ per kg mole of injectant (Section 11.4)
Carbon from lump coke in charge	Pure carbon			$n_C^{coke} = ?$
Heat demand of wustite reduction zone (Chapter 12)				$D^{wrz} = 370\,000$ (kJ per kg mole of product Fe, excluding injectant heat demands)

These two equations are simplified by the operating conditions outlined in Table 14.1 ($x = 1$, $E^B = 0$, $y = 0$) so that they may be combined to give:

$$370\,000 + n^I \cdot (16\,000) = 198\,000 \cdot (n_C^{coke} + n^I - 0.25)$$

or

$$198\,000 \cdot n_C^{coke} + 182\,000 \cdot n^I = 420\,000 \qquad (14.1)$$

where

n_C^{coke} = kg moles of C from lump coke, per kg mole of Fe,

n^I = kg moles of C from tuyère-injected fine coke, per kg mole of Fe ($x = 1$).

Equation (14.1) describes the combinations of lump and fine coke which are acceptable to the furnace. It defines, therefore, the fuel combinations from which the optimum must be chosen.

The total cost of the fuels going into the furnace is, of course, given by

$$\begin{matrix}\text{Fuel cost \$}\\ \text{per kg mole}\\ \text{of product Fe}\end{matrix} = n_C^{coke} \cdot (1.20) + n^I \cdot (0.30). \qquad (14.2)$$

It is this cost which must be minimized.

Equations (14.1) and (14.2) demonstrate the general nature of blast furnace optimization problems. They show:

(a) that there are many combinations of raw materials and operating conditions which will satisfy the stoichiometric and thermal requirements of the process (as exemplified by two unknowns in the one operating equation, (14.1));

(b) that only when a particular function is to be optimized (in this case fuel cost) can a rational choice of raw materials and operating conditions be made.

14.2 Graphical Representation of Cost Minimization

The simple optimization problem outlined in Section 14.1 is solved graphically in Fig. 14.1. Graphical solutions are limited to problems with

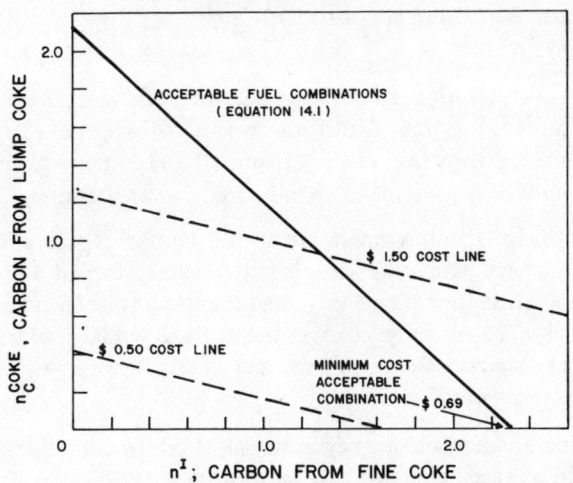

Fig. 14.1. Optimization plot for the Section 14.2 cost minimization problem showing: (i) operating line 14.1 and (ii) cost lines 14.2. The minimum cost of an acceptable fuel combination is shown to be $0.69 per kg mole of product Fe. The units of n_C^{coke} and n^I are kg moles of C per kg mole of Fe from lump coke and fine coke respectively.

only two or three variables, but they illustrate the general type of search procedures which are followed by linear programmes.

Figure 14.1 plots:

(a) equation (14.1);
(b) the family of curves described by equation (14.2) for different total fuel costs.

As the graph shows, each point of intersection of a cost line with the equation (14.1) line indicates a combination of fuels which (i) satisfies equation (14.1) and (ii) has the total cost indicated by the intersecting cost line. In this particular case the minimum cost combination is

$$\frac{\text{Minimum fuel cost}}{\text{per kg mole of Fe}} = \$0.69 \qquad n_C^{coke} = 0 \qquad n^I = 2.31$$

which is equivalent to a total fuel cost of $12.35 per tonne of Fe.

14.2.1 Inclusion of Physical Constraints in the Optimization Problem

The Fig. 14.1 solution gives the answer that the most economic combination of fuels for this blast furnace is zero lump coke and all fine (tuyère-injected) coke. However, as Chapters 2 and 3 showed, this combination would not be acceptable in an industrial furnace because:

(a) the tuyère flame temperature would be unacceptably low due to the injected fine coke entering the tuyère zone at a much lower temperature than the descending incandescent lump coke;
(b) the absence of lump coke in the furnace would result in unsatisfactory support for the charge and unsatisfactory permeability for gases rising in the shaft.

These considerations place two new constraints on the combinations of fuel which would be acceptable in an actual furnace.

14.2.2 Flame Temperature Constraint

The minimum flame temperature constraint may, for the purposes of this simplified example, be described by expressions of the type:

$$n_C^{coke} \geqslant F \cdot n^I$$ (F = a numerical coefficient which is set by the minimum acceptable flame temperature)

or

$$n_C^{coke} - F \cdot n^I \geqslant 0.$$

For example, the furnace operator may conclude that for an acceptable flame temperature, the amount of carbon from lump coke must equal or exceed 5 times the amount of carbon from fine coke, per tonne of product metal. In arithmetic terms this requirement is represented by

$$n_C^{coke} \geqslant 5 \cdot n^I$$

or

$$n_C^{coke} - 5 \cdot n^I \geqslant 0. \tag{14.3}$$

Figure 14.2 demonstrates how this additional constraint can be superimposed on Fig. 14.1. It shows that for a fuel combination to be com-

Blast-Furnace Optimization

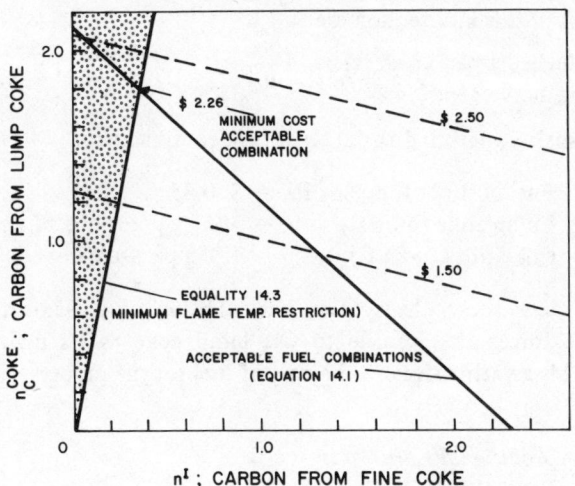

Fig. 14.2. Optimization plot showing the effect of flame temperature restriction (14.3) on minimum fuel cost. Fuel combinations acceptable to the process are now restricted to the portion of the equation (14.1) line to the left of flame temperature restriction line (14.3). The minimum fuel cost is $2.26 per kg mole of product Fe.

patible with this new restriction:

(a) the combination must, as before, lie on the equation (14.1) line;
(b) it must be on or to the left of the minimum flame temperature line (14.3).

Thus, acceptable fuel combinations are now restricted to the small top-left segment of the equation (14.1) line.

The search for the minimum-cost combination consistent with these restrictions is carried out as is shown in Fig. 14.2. This figure indicates that:

(a) the $1.50 total-cost line is now nowhere simultaneously acceptable to equation (14.1) and expression (14.3);
(b) the $2.50 line is acceptable to both, only where it crosses the equation (14.1) line;
(c) the minimum cost combination is the line which meets the equation (14.1) and equality (14.3) lines at their intersection.

188   *The Iron Blast Furnace*

The details of this intersection are:

$$\frac{\text{Fuel cost per}}{\text{kg mole of Fe}} = \$2.26 \qquad n_C^{coke} = 1.79 \qquad n^I = 0.36$$

which, translated into industrial terms, are equivalent to:

Fuel cost per tonne of Fe = $40.45.
Lump coke (pure C) = 385 kg per tonne of Fe.
Coke fines (pure C) = 77 kg per tonne of Fe.

These data show clearly that the addition of a flame temperature constraint forces the furnace to use lump coke as its main source of carbon. This greatly increases the cost of fuel for the process.

14.2.3 An Additional Constraint

As mentioned in the previous section, there is a minimum lump-coke requirement for burden support and for proper gas flow through the furnace. This minimum appears to be about 320 kg of carbon per tonne of product Fe (1.5 kg moles of C per kg mole of Fe). Expressed in inequality terms this requirement takes the form:

$$n_C^{coke} \geqslant 1.5 \text{ kg moles of C per kg mole of product Fe.} \qquad (14.4)$$

This additional constraint is shown in Fig. 14.3 which indicates that for a fuel combination to be compatible with this new restriction, it must lie in the region above the horizontal line described by expression (14.4).

It can readily be seen, however, that the previous restrictions are compatible with but more severe than the (14.4) restriction, so that it does not, in fact, influence the final optimum result.*

---

*The equations and inequalities in an optimization problem must be realistic, i.e. they must permit a solution which can satisfy all the imposed restrictions. If, for example, the additional lump coke availability restriction

$n_C^{coke} \leqslant 1.0$ (i.e. only 1 kg mole of lump coke carbon is available per kg mole of Fe)

were superimposed upon the Fig. 14.2 and 14.3 problems, it would become impossible for any fuel combination (irrespective of cost) to satisfy all the conditions of the problem, i.e. there would be no feasible solution. Infeasibility is indicated in a graphical solution by the absence of any feasible area and in a computer solution by an output statement.

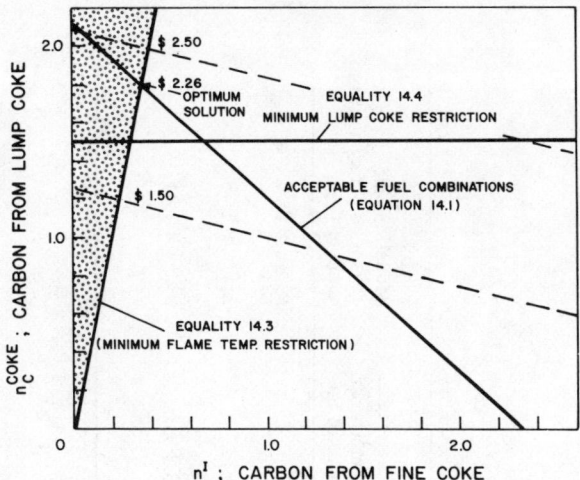

Fig. 14.3. Optimization plot showing that restriction of the lump-coke supply to not less than 320 kg per tonne of Fe does not alter the optimum determined in Fig. 14.2. The new restriction is represented by the area above the equality (14.4) line.

14.2.4 Final Operating Requirements for the Optimum Fuel-cost Solution

Once the optimum fuel combination has been chosen, the remainder of the operating parameters can be calculated. In the present case the only undetermined operating parameter is blast volume which can be determined in terms of n_O^B by means of equations (11.1) and (12.2). The overall operating conditions for the optimum fuel combination are as shown in Table 14.2.

14.3 Analytical Optimization Methods

Although graphical optimization methods are useful for demonstration purposes, industrial problems are always solved by computer techniques. A detailed description of the method is beyond the scope of this text, but a brief outline will permit an understanding of what a linear programme instructs the computer to do and what its output statements mean.

TABLE 14.2

Item	Specification	Quantity		Model variable (kg moles per kg mole of product Fe)
		kg per tonne of Fe	kg moles per tonne of Fe	
Injectant	Coke fines 298 K (pure carbon)	77		$n^I = 0.36$
Carbon from lump coke in charge	Pure carbon	385		$n_C^{coke} = 1.79$
Oxygen from dry blast air		405 (1350 Nm³ of dry air)	25.2 (O)	$n_O^B = 1.41$

Blast-Furnace Optimization

The most common computational technique is that provided by the Simplex Algorithm or a modification thereof. An important part of the Simplex method has already been demonstrated by the graphical examples above, i.e. that the optimum solution always occurs at a vertex (extreme point) on the boundary of a feasible region. In the two dimensional case described in Fig. 14.2, for example, the optimum combination occurs at the vertex described by the intersection of the equation (14.1) line and the equality (14.3) boundary line. Similarly, the Fig. 14.1 example shows its optimum combination to occur at the vertex of the equation (14.1) line and the tacitly assumed:

$$n_C^{coke} \geqslant 0$$

boundary line.

The Simplex Algorithm acknowledges the fact that the optimum solution is always at a vertex. It (i) examines the vertices, (ii) rejects those which do not satisfy all the equations and physical restrictions and (iii) determines which of the feasible vertex solutions gives the optimum result.

14.4 Computer Inputs and Outputs

This section illustrates, re-using the Section 14.2.2 problem as an example, a typical interactive computer optimization operation. The programme is shown in Table 14.3. After the size of the problem* and the objective (maximization or minimization) have been ascertained, the computer requests the form of the objective function. In this case the objective function equation is:

$$1.2 \cdot n_C^{coke} + 0.3 \cdot n_C^I = \text{Fuel Cost} \qquad (14.2)$$

which is entered into the computer by means of its coefficients 1.2 and 0.3.

---

*There is no general rule with regard to the number of constraints (equations plus inequalities) and the number of unknowns. There need only be a feasible solution for the technique to work.

TABLE 14.3

Input Sequence for the Interactive Linear Optimization Programme LINPRG as applied to the Section 14.4 example problem
LINPRG is a conversational mode linear programme developed for the McGill University interactive computing system. Many computing systems have similar programmes as part of their software

```
/exec linprg
*IN PROGRESS


ENTER THE NUMBER OF VARIABLES AND THE NUMBER
OF CONSTRAINTS
TYPE * OR $ TO INPUT FROM DISK
?
         2                  2

TO MAXIMIZE THE OBJECT FUNCTION, TYPE +1.0
TO MINIMIZE THE FUNCTION, TYPE -1.0
TYPE * OR $ TO INPUT FROM DISK
?
-1.0

TO PRINT THE FIRST AND LAST TABLEAU ONLY, TYPE 1
TO PRINT EVERY TABLEAU, TYPE 2
TYPE * OR $ TO INPUT FROM DISK
?
1

ENTER THE COEFFICIENTS OF THE OBJECTIVE FUNCTION
TYPE * OR $ TO INPUT FROM DISK
?
         1.2                0.3

IN TRIPLETS, ENTER ROW NO., COLUMN NO., VALUE FOR
EACH NON-ZERO ELEMENT OF L.H.S.
TYPE ZERO AFTER LAST TRIPLET
TYPE * OR $ TO INPUT FROM DISK
?
1 1   198000           1 2   182000
?
2 1      1             2 2      -5
?
0

ENTER THE TYPE OF CONSTRAINTS OF RESPECTIVE EQUATIONS
1 SIGNIFIES LESS THAN OR EQUAL,2 SIGNIFIES EQUAL,
3 SIGNIFIES GREATER THAN OR EQUAL; USE ONE LINE TO ENTER DATA.
TYPE * OR $ TO INPUT FROM DISK
?
         2                  3

ENTER THE R.H.S. (POSITIVE NUMBER) OF THE RESPECTIVE EQUATIONS.
TYPE * OR $ TO INPUT FROM DISK
?
      420000                0.0
```

Blast-Furnace Optimization

The operating equations and physical restriction expressions are then entered into the computer in matrix form by:

(a) representing the left-hand sides of the expressions:

$$198\,000 \cdot n_C^{coke} + 182\,000 \cdot n^I = 420\,000 \tag{14.1}$$

$$n_C^{coke} - 5 \cdot n^I \geq 0.0 \tag{14.3}$$

in the coefficient form:

1	1	198 000	1	2	182 000
2	1	1	2	2	−5

(b) by indicating that the first expression is an equation and the second a greater-than-or-equal-to statement;
(c) by indicating that the right-hand sides of equation (14.1) and expression (14.3) are 420 000 and 0.0 respectively.

Of course, for a larger problem this information is much more efficiently entered into the computer on cards or tape.

14.4.1 Computer Print Out

The computer prints out the results of its optimization search in the form of a series of tableaux. The initial and final tableaux (Table 14.4) describe the problem and the optimum solution respectively, and hence they are the most significant parts of the output.

The most noticeable feature of the initial tableau (A) is that the Simplex Algorithm introduces 'slack' variables (S_2) and 'artificial variables' (a_1, a_2) into the system of equalities and inequalities. These variables (subscripted by their equation number) are introduced in order to (i) express and solve all the equalities and inequalities in terms of equations and to (ii) identify the expressions as the $\leq$, $=$ or $\geq$ type.

The identifications are:

$\leq$ a slack variable S
$=$ an artificial variable a
$\geq$ $-S + a$

In determining the optimum solution, the Simplex Algorithm:

(a) forces all artificial variables to zero (which they must be, for example, in all equality expressions);
(b) determines the value of each slack variable under the optimum solution conditions. This value of the slack variable indicates the magnitude by which each $\leqslant$ or $\geqslant$ expression deviates from its equality component. In Table 14.4, for example, slack variable S_2 is zero in the optimum solution (Tableau B) because the optimum lies (Fig. 14.2) on the $n_C^{coke} - 5 \cdot n^I = 0$ line.*

These mathematical procedures lead to a final optimum tableau such as that shown in Table 14.4B. The most important parts of the final tableau are:

(a) 'value' which indicates the optimum value of the objective function, in this case the minimum total fuel cost ($2.26 per kg mole of product Fe);
(b) 'Solution' which indicates the optimum combination of variables, in this case the optimum combination of lump coke and tuyère-injected fine coke (1.79 and 0.36 kg moles of C per kg mole of product Fe respectively).

The solution also indicates zero values for the artificial variables and in this particular case a zero value for the slack variable (S_2) in expression (14.3).

The final tableau usually gives additional information with regard to improving the optimum. In the present case this might take the form of lowering the optimum fuel cost by relaxing the flame-temperature constraint (expression (14.3)) or by lowering the heat demand of the process (equation (14.1)). These details are too complex to be included in this text but they are discussed at length in the references which are noted at the end of the chapter. Much of this additional information can also be obtained by repeating the above optimization procedures with altered numerical data.

---

*If, for example, restriction 14.4 is included in the calculation, its slack variable has the value of 0.29 under the optimum solution conditions. This means that when the optimum combination of n_C^{coke} and n^I is inserted into expression (14.4), the left-hand side exceeds the right-hand side by 0.29 kg moles of lump coke carbon per kg mole of product Fe.

Blast-Furnace Optimization

Interpretation of Print-out from Linear Optimization Programme LINPRG as applied to the Section 14.4 Example Problem: (A) Initial Tableau Describing the Problem; (B) Final (Optimum) Tableau Showing the Optimum Value of the Objective Function and the Optimum Fuel Combination

A. Initial Tableau

Coefficients					Value = 10^{30}		Meanings
1	2	3	4	5	Right-hand side		Value = \$$10^{30}$ (artificially high value set up by method of eliminating artificial variables)
198 000	182 000	1	0	0	420 000		$198\,000 \cdot n_C^{coke} + 182\,000 \cdot n^I + 1.0 \cdot a_1 = 420\,000$
1	−5	0	−1	1	0		$1.0 \cdot n_C^{coke} - 5.0 \cdot n^I - 1.0 \cdot S_2 + 1.0 \cdot a_2 = 0$
−1.2	−0.3	0	0	0	0		$1.2 \cdot n_C^{coke} + 0.3 \cdot n_C^{coke} =$ Fuel cost

(the negative coefficients on 1.2 and 0.3 in the initial tableau indicate that the fuel cost is to be minimized)

B. Final **Optimum** Tableau

Value = 2.26

Minimum = 2.26 \$ per kg mole fuel cost = of product Fe

Data for optimum combination (kg moles of C per kg mole of Fe)

	n_C^{coke}	n^I	a_1	S_2	a_2
Solution	1.79	0.36	0	0	0
			driven to zero	zero in this case	driven to zero

1	2	3	4	5
1.79	0.36	0	0	0

14.5 A More Complete Problem

The example problems above, though illustrative, greatly oversimplify actual industrial situations, i.e. they do not represent the wide range of raw materials and operating conditions which are available to the blast-furnace operator. For this reason, the next few sections present a more complex problem to indicate how a large number of variables and a large number of constraints (operating equations plus physical restriction expressions) can be incorporated into a linear optimization programme.

For this more complex case, it will be assumed that the blast-furnace operator has at his disposal the raw materials described in Table 14.5. The operator's problem is to ascertain which combination of raw materials, acceptable to the process, will result in a minimum total raw materials cost per tonne of Fe. The specifications for the problem are as shown in Table 14.6. In addition it is specified that:

(a) the weight of coke must not be less than 360 kg per tonne of product (Girard, 1971) Fe for proper charge support and gas permeability;
(b) the weight of injected oil must not exceed 0.15 kg per Nm^3 of dry blast air (to ensure complete combustion);
(c) the tuyère flame temperature, as defined in Appendix I, must not be less than 2200 K or greater than 2600 K, ignoring coke ash.

14.5.1 The Objective Function

The total cost of raw materials for producing 1 kg mole of Fe, the item to be minimized, may, from Table 14.5, be expressed as

$$n_C^{coke} \cdot (1.33) + n_{CaCO_3} \cdot (1.00) + W_{oil}^I \cdot (0.09) + n_{O_2}^I \cdot (0.64)$$
$$+ n_{Fe}^i \cdot (1.68) = \text{total raw materials cost, \$ per kg mole of product Fe.}$$
(14.5)

It will be immediately noticed that certain model variables (e.g. n_O^B, n_C^A, D^{wrz}) do not appear in Cost Equation (14.5). This is because:

(a) they have no costs associated with them in this particular problem (e.g. n_O^B); or
(b) their cost is already represented by other variables in the cost

TABLE 14.5
Blast-furnace Raw Materials and Their Prices (1978)

Raw material	Price (industrial units)	Price (equation units)
Coke (90% C, 10% SiO_2)	$100 (tonne)$^{-1}$	$1.33 (kg mole of C)$^{-1}$
Iron ore (95% Fe_2O_3, 5% SiO_2)	$30 (tonne of contained Fe)$^{-1}$	$1.68 (kg mole of Fe)$^{-1}$
Limestone (100% $CaCO_3$)	$10 (tonne)$^{-1}$	$1.00 (kg mole of $CaCO_3$)$^{-1}$
Oil* (86 wt.% C, 14% H)	$13 (barrel)$^{-1}$	$0.09 (kg of oil)$^{-1}$
Oxygen	$20 (tonne)$^{-1}$	$0.64 (kg mole of O_2)$^{-1}$

*1 barrel of oil = 0.16 m^3. Density of oil $\approx$ 900 kg m^{-3}.

equation. An example missing variable of this type is n_C^A (active carbon), the cost of which is represented in the costs of n_C^{coke}, W_{oil}^I and n_{CaCO_3}.

These zero cost variables* are entered in the objective equation by means of 0.0 cost coefficients.

14.5.2 Operating Equations

The chemical and thermal requirements of the blast furnace are represented in the linear programme by equations of the type:

$$n_{Fe}^i = n_{Fe}^o = 1, \tag{4.4}$$

$$n_O^B + 1.06 + z \cdot n^I + 2(Si/Fe)^m + 2n_{CaCO_3}$$
$$= 1.3 \, n_C^A + 0.38 \cdot y \cdot n^I, \tag{12.3}$$

$$n_C^A + (C/Fe)^m = n_C^{coke} + x \cdot n^I + n_{CaCO_3}, \tag{12.2}$$

$$D^{wrz} + n^I \cdot D^I = S^{wrz}$$
$$= 198\,000 \cdot n_C^A + E^B \cdot n_O^B + 95\,000 \cdot y \cdot n^I. \tag{11.12}$$

*All variables which need not be included in the objective function (the item being maximized or minimized) are given 0.0 coefficients. If, for example, coke requirement per tonne of product Fe were being minimized rather than raw materials cost, only n_C^{coke} would appear in the objective equation. All other variables would have 0.0 coefficients.

TABLE 14.6

Item	Specification	Quantity kg per tonne of Fe	Quantity kg moles per tonne of Fe	Model variable	
Ore	Hematite (5%SiO$_2$)	1000			
Fe			17.9		
Iron oxide entering wustite reduction zone	Fe$_{0.947}$O			(O/Fe)$^{x\text{-wrz}}$	= 1.06
Pig iron	1% Si, 5% C	Carbon 53	4.4	(C/Fe)^m	= 0.25
		Silicon 11	0.4	(Si/Fe)^m	= 0.021
Coke	90% C, 10% SiO$_2$ dry basis			n_C^{coke}	= ?
Blast	8 g of H$_2$O per Nm^3 moist blast			n_O^B	= ?
				$n_{H_2O}^B$	= 0.024 n_O^B
Blast temperature				T_B	= 1300 K
Injectants					
(a) pure O$_2$ (T_B)				$n_{O_2}^I$	= ?
				z	= 2
(b) oil (298 K)	86 wt.% C, 14% H			wt·$_{oil}^I$	= ?
				H^f_{298} oil	= −1660 per kg of oil
				D^I_{oil}	= 4700 kJ per kg of oil
Flux	CaCO$_3$			n_{CaCO_3}	= ?
Slag	CaO/SiO$_2$ weight ratio = 1.2			wt·$_{slag}$	= ?
Convective and				D^{loss}	= 20 000

Blast-Furnace Optimization

There are three injectants (oil, oxygen, blast moisture) in this particular problem so that the actual equations in the programme will be somewhat more complex than those shown here. One injectant equation which is immediately available, however, is the relationship between the quantity of moisture injected in the blast (8 g per Nm³ of moist blast) and n_O^B, i.e.

$$n_{H_2O}^I = 0.024 \, n_O^B. \tag{14.6}$$

(See Problems 3.1c, 12.1)

14.5.3 Additional Operating Equations

Specification by the problem that the slag must have a CaO/SiO_2 weight ratio of 1.2 furnishes an additional operating equation for the process, i.e.

$$\text{wt.}_{CaO \text{ in slag}} = 1.2 \cdot \text{wt.}_{SiO_2 \text{ in slag}}$$

or

$$\text{wt.}_{CaO \text{ in } CaCO_3}$$
$$= 1.2 \cdot (\text{wt.}_{SiO_2 \text{ in ore}} + \text{wt.}_{SiO_2 \text{ in coke}} - \text{wt.}_{SiO_2 \text{ reduced to Si}}) \tag{14.7}$$

which is straightforwardly related to the variables n_{CaCO_3}, n_C^{coke}, $(Si/Fe)^m$ and the composition of the iron ore. Furthermore, slag weight is represented (assuming the slag to be made up of CaO and SiO_2 only, i.e. with no Al_2O_3 or MgO) by

$$\text{wt.}_{slag} = \text{wt.}_{CaO \text{ in slag}} + \text{wt.}_{SiO_2 \text{ in slag}}$$

or

$$\text{wt.}_{slag} = 2.2 \cdot (\text{wt.}_{SiO_2 \text{ in ore}} + \text{wt.}_{SiO_2 \text{ in coke}} - \text{wt.}_{SiO_2 \text{ reduced to Si}}). \tag{14.8}$$

Lastly, the limestone and slag weight specifications by equations (14.7) and (14.8) furnish the final information (n_{CaCO_3}, wt._{slag}) required for calculating wustite reduction zone heat demand, i.e.

$$D^{wrz} = D^{Fe} + (C/Fe)^m \cdot D^C + D^{loss} + (Si/Fe)^m \cdot D^{Si}$$
$$+ \text{wt.}_{slag} \cdot D^{slag} + n_{CaCO_3} \cdot D^{CaCO_3}. \tag{12.13}$$

Procedures for calculating the numerical values in equation (12.13) are presented in Chapter 12.

14.5.4 Physical Restrictions

The final constraint expressions to be entered into the linear programme are those provided by the physical requirements of the blast furnace. The minimum coke restriction (360 kg of coke per tonne of Fe, 90% C, Girard, 1971) is inserted into the programme in the form:

$$n_C^{coke} \geqslant 1.5 \tag{14.9}$$

while the maximum oil loading of 0.15 kg per Nm³ of dry blast air is represented by

$$W_{oil}^I \leqslant 0.15 \cdot \{\text{volume of dry blast air (Nm}^3\text{)}\}$$

$$\leqslant 0.15 \cdot \left\{\frac{n_O^B}{2} \cdot \frac{22.4}{0.21}\right\}$$

or

$$W_{oil}^I - n_O^B \cdot 8.0 \leqslant 0. \tag{14.10}$$

Flame temperature restrictions are inserted into the optimization programme in the form of expressions developed as shown in Appendix I. The expressions are of the type:

(a) $T_F \geqslant 2200$ K,

$$A \cdot n_O^B + B \cdot n_{H_2O}^I + C \cdot n_{O_2}^I + D \cdot W_{oil}^I \geqslant 0; \tag{14.11}$$

(b) $T_F \leqslant 2600$ K,

$$E \cdot n_O^B + F \cdot n_{H_2O}^I + G \cdot n_{O_2}^I + H \cdot W_{oil}^I \leqslant 0 \tag{14.12}$$

where A, B, C, D, E, F, G and H are constants determined by the prescribed flame temperatures and the temperature of the blast. The constants for this particular problem are given in Appendix AI.1 (pp. 209, 210).

14.5.5 Running the Programme

The programme is run exactly as shown in Section 14.4. The number of variables:

10: $n_{Fe}^i, n_O^B, n_{H_2O}^I, n_{O_2}^I, n_C^A, n_C^{coke}, W_{oil}^I, n_{CaCO_3}, \text{wt.}_{slag}, D^{wrz}$

and the number of restrictions (12: expressions (4.4), (11.12), (12.2), (12.3), (12.13), (14.6), (14.7), (14.8), (14.9), (14.10), (14.11) (14.12)) are entered; the objective and descriptive (constraint) equations are described; and the search is set in motion.

In this particular problem the optimum value is:

$$\text{minimum total raw material cost per kg mole of product Fe} = \$4.75$$

(\$85.0 per tonne of Fe) and the solution is:

n_O^B = 1.54 (kg moles per kg mole of product Fe)

$n_{O_2}^I$ = 0.0 (kg moles per kg mole of product Fe)

n_{CaCO_3} = 0.115 (kg moles per kg mole of product Fe)

n_C^{coke} = 1.83 (kg moles per kg mole of product Fe)

W_{oil}^I = 5.7 (kg per kg mole of product Fe).

The final operational table for this minimum cost operation in model and industrial units is shown in Table 14.7.

14.5.6 Interpretation

The results of this particular optimization programme show that the minimum raw materials cost is realized when as much oil as possible (without causing the flame temperature to fall below 2200 K) is injected into the furnace. The results also show that at the prices of this example (Table 14.5), pure oxygen cannot be afforded as a means of offsetting the cooling effect of the oil, i.e. the use of oxygen to increase the permissible oil-injection level is not economically favoured.

TABLE 14.7

Item	Quantity (kg per tonne of Fe)	Model variable (kg moles per kg mole of product Fe)	
Ore (95% Fe_2O_3)	1505		
Oxygen from dry blast air	442 (1475 Nm^3 of dry air, 1490 Nm^3 of moist air)	n_O^B	= 1.54
Coke (90% C)	393 (C) 437 (coke)	n_C^{coke}	= 1.83
Oil	102	W_{oil}^I	= 5.7 [kg]
Oxygen	0	$n_{O_2}^I$	= 0.0
Limestone	206	n_{CaCO_3}	= 0.115
Slag	211	W_{slag}	= 11.8 [kg]
Flame temperature (ignoring coke ash)		2200 K	

Of course, should the price of oil increase or the price of coke decrease, there could be a turnabout towards higher coke levels. The optimum raw materials combination under such new cost conditions can easily be ascertained by re-running the programme with an updated cost equation (Problem 14.3).

14.6 Summary

This chapter has shown how linear programming techniques, using the equations of this text and known physical constraints on industrial furnaces, can be used to predict how a blast furnace should be run to achieve a specified optimization goal.

The requirements of linear programming:

(a) that the operating variables of the process be related to each other in a linear manner by valid operating and physical constraint expressions;
(b) that the objective (e.g. cost) function be related linearly to the operating variables;

Blast-Furnace Optimization

are met by the blast furnace. It is therefore readily amenable to optimization calculations.

Suggested Reading

Althoen, S. C. and Bumcroft, R. J. (1976) *Matrix Methods in Finite Mathematics, An Introduction with Applications to Business and Industry*, W. W. Norton & Co., New York.

Strum, J. E. (1972) *Introduction to Linear Programming*, Holden-Day Inc., San Francisco.

Reference

Girard, M., Jusseau, N. and Michard, J. (1971) 'Possible developments in blast furnace practice – prospective study using an operating model', in *Proceedings of International Conference on The Science and Technology of Iron and Steel*, Part 1, The Iron and Steel Institute of Japan, Tokyo, p. 117.

Chapter 14 Problems. *Optimization by Linear Programming*

14.1 The research department of the company operating the Section 14.1/14.2 blast furnace suggests that the minimum lump-coke (pure C) requirement should, for improved gas distribution, be raised to:

$$n_C^{coke} \geq 1.9.$$

Determine what the minimum fuel cost would be if this proposed restriction were to be imposed. Use an analytical linear programme.

14.2 It is suggested that with changing economic conditions, the Section 14.1/14.2 blast furnace should be injected with CH_4 rather than coke fines. Given forecasts that the prices of coke and CH_4 will be

	Coke (pure C)	CH_4
(a)	$120 per tonne	$50 per 1000 Nm³
(b)	$120 per tonne	$100 per 1000 Nm³,

advise the operator as to what combinations of lump coke and CH_4 should be employed in order to minimize his total fuel cost. State for each case the quantities of fuels required and the total minimum fuel cost, per tonne of Fe. The coke quantity constraint for the furnace is

$$n_C^{coke} \geq 1.5$$

and the flame temperature constraint is

$$\begin{Bmatrix} \text{wt. of coke} \\ \text{(pure C, kg)} \end{Bmatrix} \geq 2 \cdot \begin{Bmatrix} \text{volume of } CH_4 \\ (Nm)^3 \end{Bmatrix}$$

14.3 The operator of the blast furnace in Section 14.5 wishes to examine the manner in which oil price changes would affect his operations. Determine for him the minimum total raw materials cost ($ per tonne of Fe) which would have to be paid if the price of oil were to increase to $15, $20 and $25 per barrel.

14.4 The research department of the Section 14.5 blast-furnace plant suggests that the minimum flame temperature restriction should be raised to 2400 K. What effect would this have on the minimum raw materials cost (oil @ $13 per barrel) and what would the optimum combination of raw materials for this suggested new operation be?

14.5 The steel plant associated with the Section 14.5 blast furnace demands a maximum hot metal supply irrespective of cost. Suggest to the operator how he might meet this maximum demand in terms of the amounts of oil and oxygen he should inject. Assume that the hot metal production rate is (i) inversely proportional to the total volume (Nm^3) of moist blast air + pure O_2 being consumed by the furnace per tonne of Fe; and (ii) independent of the amount of oil being injected. Continue with the restriction that the minimum flame temperature is 2400 K and introduce the oxygen availability limitation:

$$\frac{\text{volume of pure injected } O_2 \text{ (Nm}^3)}{} \leqslant 0.05 \cdot \left\{ \begin{array}{l} \text{volume of moist} \\ \text{blast air (Nm}^3) \end{array} \right\}$$

APPENDIX I

Tuyère Flame Temperature Calculations

Tuyère flame temperature, as defined in this text, is the temperature which would be attained by the combustion products from burning:

   (a) incandescent coke (1800 K);
   (b) hydrocarbon tuyère injectants;

with:

   (c) high temperature blast air at its temperature of entry into the furnace (T_B);
   (d) tuyère-injected oxygen (also at T_B);

to form carbon monoxide and hydrogen. It is assumed that there are no heat losses from the flame, i.e. that the combustion is adiabatic.

In physical terms, the tuyère flame temperature defined here closely represents conditions just outside the coke percolator at the edge of the tuyère raceways, i.e. at the point where the carbon, oxygen and hydrogen have just formed CO and H_2.

The basic presumptions of flame temperature calculations are that:

   (a) hydrocarbons entering the furnace through the tuyères consume as much tuyère oxygen (from air, moisture and injected O_2) as is necessary for them to form CO and H_2;
   (b) the remaining tuyère oxygen is used to combust incandescent coke to CO. The coke is assumed to have attained a temperature of 1800 K during its descent.

The calculation procedures are illustrated by the following two examples.

(a) Sample Calculation 1. Coke combustion only: no tuyère-injected hydrocarbons

With no hydrocarbon injection through the tuyères, all the oxygen of the blast air, blast moisture and injected oxygen goes to the combustion of coke. The flame temperature for this simplified case is calculated as follows:

Example operating conditions:

Blast temperature	T_B = 1300 K
Blast moisture	7.5 g H_2O/Nm³· of dry blast air (0.022 kg moles of H_2O/kg mole of O from dry blast air)
Injected oxygen (pure O_2)	0.03 Nm³ of O_2/Nm³ of dry blast air (0.071 kg moles of O_2/kg mole of O from dry blast air)
Injected hydrocarbons	0.

Basis of calculation: 1 kg mole of O from dry blast air:

Reactants	Products (T_F)
1 kg mole of O from dry air (1300 K)	1 kg mole of CO
0.022 kg mole of H_2O (1300 K)	{0.022 kg mole of CO} {0.022 kg mole of H_2}
0.071 kg mole of O_2 (1300 K)	0.142 kg mole of CO
1.164 kg moles of C from coke (1800 K)	{included in CO formed with air, H_2O and pure O_2}
$\frac{1}{2} \cdot \frac{0.79}{0.21}$ kg mole of N_2 (1300 K)	1.88 kg moles of N_2

Enthalpy in Reactants = Enthalpy in Products

$$\frac{1}{2} \cdot H^\circ_{1300 \atop O_2} + 0.022 \cdot H^\circ_{1300 \atop H_2O} + 0.071 \cdot H^\circ_{1300 \atop O_2} + 1.164 \cdot H^\circ_{1800 \atop C} + 1.88 \cdot H^\circ_{1300 \atop N_2}$$

$$= 1.164 \cdot H^\circ_{T_F \atop CO} + 0.022 \cdot H^\circ_{T_F \atop H_2} + 1.88 \cdot H^\circ_{T_F \atop N_2}$$

Enthalpy data (kJ per kg mole, Appendices V & VI)

$H^\circ_{1300 \atop O_2}$ = 33 600; $\quad H^\circ_{1300 \atop H_2O}$ = −242 000 + [39 400]; $\quad H^\circ_{1800 \atop C}$ = 30 500

$H^\circ_{1300 \atop N_2}$ = 31 700; $\quad H^\circ_{T_F \atop CO}$ = −111 000 + [36.6 T_F − 16 400]

$H^\circ_{T_F \atop H_2}$ = [35.2T_F − 17 200]; $\quad H^\circ_{T_F \atop N_2}$ = [36.4T_F − 16 500]

Flame temperature (calculated):
2590 K

Tuyère Flame Temperature Calculations

(b) Sample Calculation 2. Conditions as in sample calculation 1 but with the injection of 0.1 kg of oil per Nm^3 of dry blast air

This example shows how tuyère injectants are included in flame-temperature calculations.

Operating conditions

Blast temperature	$T_B = 1300$ K
Blast moisture	7.5 g/Nm^3 of dry blast air
Injected oxygen	0.03 Nm^3/Nm^3 of dry blast air
Injected hydrocarbon	0.1 kg of oil (86% C, 14% H)/Nm^3 of dry blast air (5.33 kg of oil or 0.38 kg mole of C + 0.37 kg mole of H_2 per kg mole of O from dry blast air)

Basis of calculation: 1 kg mole of O from dry blast air:

Reactants	Products (T_F)
1 kg mole of O from dry air (1300 K)	1 kg mole of CO
0.022 kg mole of H_2O (1300 K)	{0.022 kg mole of CO 0.022 kg mole of H_2}
0.071 kg mole of pure O_2 (1300 K)	0.142 kg mole of CO
0.38 kg mole of C from oil (298 K) (enthalpy included in oil enthalpy)	{included in CO formed with air, H_2O and pure O_2}
0.784 kg mole of C from coke (1800 K) burnt by the O from air, H_2O and pure O_2	{included in CO formed with air, H_2O and pure O_2}
which is left over from combusting oil	
0.37 kg mole of H_2 from oil (298 K) (enthalpy included in oil enthalpy)	0.37 kg mole of H_2
1.88 kg moles of N_2 from dry air (1300 K)	1.88 kg moles of N_2

Enthalpy in Reactants = Enthalpy in Products

$$\tfrac{1}{2} \cdot \underset{O_2}{H^°_{1300}} + 0.022 \cdot \underset{H_2O}{H^°_{1300}} + 0.071 \cdot \underset{O_2}{H^°_{1300}} + 5.33 \cdot \underset{\substack{298 \\ \text{oil}}}{H^°_W} + 0.784 \cdot \underset{C}{H^°_{1800}}$$

$$+ 1.88 \cdot \underset{N_2}{H^°_{1300}} = 1.164 \cdot \underset{CO}{H^°_{T_F}} + 0.392 \cdot \underset{H_2}{H^°_{T_F}} + 1.88 \cdot \underset{N_2}{H^°_{T_F}}$$

The Iron Blast Furnace

Additional enthalpy data:

$H^\circ_{W_{298}^{oil}}$ = enthalpy of oil (298 K) $\boxed{\text{kJ per kg of oil}}$ (see note C, Section AI.2)

= -1660 kJ per kg of oil (in this case)

Flame temperature (calculated):
  2200 K

The flame temperature in this case is much lower than that in Sample Problem 1 (2590 K) which confirms that hydrocarbon injection through the tuyères tends to cool the tuyère zone.

AI.1 Flame-Temperature Equations for Linear Programme Constraints

Equations relating flame temperature to the operating variables of Chapter 4 onwards are necessary (i) to check that the flame temperature from a proposed set of operating parameters is acceptable to the furnace and (ii) to provide constraint equations for blast-furnace optimization programmes (Chapter 14).

Such equations may be developed by the procedures demonstrated in Sample Calculation 2, above. Referring back to this example, the calculation is put on the new basis of:

n^B_O kg moles of O from dry blast air per kg mole of product Fe

$n^I_{H_2O}$ kg moles of H_2O from blast moisture per kg mole of product Fe

$n^I_{O_2}$ kg moles of pure O_2 injected with blast (T_B) per kg mole of product Fe

W^I_{oil} kg of oil (86% C, 14% H) injected with blast per kg mole of product Fe

(which provide $\dfrac{86}{100} \cdot \dfrac{1}{12} \cdot W^I_{oil}$ and $\dfrac{14}{100} \cdot \dfrac{1}{2} \cdot W^I_{oil}$ kg moles of C and H_2, respectively).

Tuyère Flame Temperature Calculations

Following through Sample Calculation 2 above with these new quantities leads to the equation (T_B = blast temperature):

$$n_O^B \cdot \tfrac{1}{2} \cdot H_{T_B \, O_2}^\circ + n_{H_2O}^I \cdot H_{T_B \, H_2O}^\circ + n_{O_2}^I \cdot H_{T_B \, O_2}^\circ + W_{oil}^I \cdot H_{W \, 298 \, oil}^\circ$$

$$+ \left\{ (n_O^B + n_{H_2O}^I + 2 \cdot n_{O_2}^I) - \frac{\text{wt.\% C}}{1200} \cdot W_{oil}^I \right\} \cdot H_{1800 \, C}^\circ$$

$$+ 1.88 \cdot n_O^B \cdot H_{T_B \, N_2}^\circ = (n_O^B + n_{H_2O}^I + 2 \cdot n_{O_2}^I) \cdot H_{T_F \, CO}^\circ$$

$$+ \left(n_{H_2O}^I + \frac{\text{wt.\% H}}{200} \cdot W_{oil}^I \right) \cdot H_{T_F \, H_2}^\circ + 1.88 \cdot n_O^B \cdot H_{T_F \, N_2}^\circ. \qquad (AI.1)$$

This equation is of a form which can be used generally, making sure, however, that the prescribed blast temperature (1300 K in this example) and injectant compositions and temperatures are inserted.

In this particular case, incorporation of the appropriate oil composition and enthalpy data from the sample problems above leads to:

$$n_O^B \cdot (265\,300) + n_{H_2O}^I \cdot (-28\,000) + n_{O_2}^I \cdot (349\,000) + W_{oil}^I \cdot (-2650)$$

$$= n_O^B \cdot (105.0 \cdot T_F) + n_{H_2O}^I \cdot (71.8 \cdot T_F) + n_{O_2}^I \cdot (73.2 \cdot T_F)$$

$$+ W_{oil}^I \cdot (2.46 \cdot T_F) \qquad (AI.2)$$

from which flame temperatures can quickly be calculated.

Flame-temperature constraint equations for linear-programming purposes may be developed by inserting the prescribed flame-temperature limit into equation (AI.2) and by stating that the left-hand side of the equation must be equal to or greater than (or less than, depending on the problem) the right-hand side of the equation. For example, if the required inequality is

$$T_F \geqslant 2200;$$

then,

$$n_O^B \cdot (265\ 300) + n_{H_2O}^I \cdot (-28\ 000) + n_{O_2}^I \cdot (349\ 000)$$
$$+ W_{oil}^I \cdot (-2650)$$
$$\geqslant \left\{ \begin{array}{l} n_O^B \cdot (105.0) \cdot (2200) + n_{H_2O}^I \cdot (71.8) \cdot (2200) \\ + n_{O_2}^I \cdot (73.2) \cdot (2200) + W_{oil}^I \cdot (2.46) \cdot (2200) \end{array} \right\}$$

which simplifies for linear-programme purposes to

$$n_O^B \cdot (34\ 300) + n_{H_2O}^I \cdot (-186\ 000) + n_{O_2}^I \cdot (188\ 000)$$
$$+ W_{oil}^I \cdot (-8060) \geqslant 0.$$

Similarly if the required inequality is

$$T_F \leqslant 2600\ K;$$

the constraint equation becomes

$$n_O^B \cdot (-7700) + n_{H_2O}^I \cdot (-215\ 000) + n_{O_2}^I \cdot (159\ 000)$$
$$+ W_{oil}^I \cdot (-9050) \leqslant 0.$$

AI.2 Additional Items in the Calculations

(a) Coke ash

Combustion of coke at the tuyères leads to the presence of ash in the combustion products. Coke ash has been ignored in the flame-temperature calculations of this text.

(b) Minor components in injectants

Minor components in hydrocarbon injectants such as N, O and S may be included in the flame-temperature calculations (S going to H_2S) but these components have little impact on T_F.

(c) Enthalpy data for complex hydrocarbons (coal, natural gas, oil)

Fundamental enthalpy data for natural hydrocarbons are often unavailable. However, they are readily determined from measured or pub-

Tuyère Flame Temperature Calculations

lished heats of combustion. For example, the heat of formation of a hydrocarbon (298 K) is related to its gross heat of combustion (*Handbook of Chemistry and Physics*, 1977, p. D279) by*:

$$H^C_W = \frac{1}{12} \cdot \frac{wt.\% \, C}{100} \cdot H^f_{298 \, CO_2} + \frac{1}{2} \cdot \frac{wt.\% \, H}{100} \cdot H^f_{298 \, H_2O \, (l)} - H^f_{W \, 298 \, hydrocarbon}$$

where H^C_W = gross heat of combustion (298 K, to liquid H_2O) of the hydrocarbon, a negative number;

$H^f_{W \, 298 \, hydrocarbon}$ = heat of formation of the hydrocarbon from elements;

both kJ per kg of hydrocarbon.

Furthermore, since the enthalpies of elements are zero at 298 K,† the enthalpy content of the hydrocarbon (the term required in equations (AI.1) and (AI.2)) is equal to $H^f_{W \, 298}$, i.e.

$$H^\circ_{W \, 298 \, hydrocarbon} = H^f_{W \, 298 \, hydrocarbon}$$

The enthalpy content of the oil considered in Sample Calculation 2 (−1660 kJ per kg of oil, 86 wt.% C, 14 wt.% H) was calculated from its published heat of combustion −46 600 kJ per kg (*Combustion Handbook* 1965).

---

*The method of determining H^f for gaseous hydrocarbons is indicated in Appendix II.1.b.

†Enthalpy contents of hydrocarbons at temperatures other than 298 K are, in the absence of other information, approximated by:

$$H^\circ_{W \, T \, hydrocarbon} = H^f_{W \, 298 \, hydrocarbon} + \frac{1}{12} \cdot \frac{wt.\% \, C}{100} \cdot H^\circ_{T_C} + \frac{1}{2} \cdot \frac{wt.\% \, H}{100} \cdot H^\circ_{T_{H_2}},$$

i.e. $H^f_{W \, hydrocarbon}$ can be reasonably assumed to be constant with temperature.

APPENDIX II

Representing Complex Tuyère Injectants in the Operating Equations

AII.1 Gaseous Injectants with Known Heats of Combustion and Chemical Compositions

(a) Stoichiometric effects

Consider a natural gas of composition:

- α vol.% CH_4,
- β vol.% C_2H_6,
- γ vol.% N_2,
- ϵ vol.% O_2.

One kg mole (22.4 Nm^3) of this gas contains:

Item	Quantity
kg moles of C per kg mole of injectant	$= \dfrac{\alpha}{100} + 2 \cdot \dfrac{\beta}{100} = x,$
kg moles of H_2 per kg mole of injectant	$= 2 \cdot \dfrac{\alpha}{100} + 3 \cdot \dfrac{\beta}{1000} = y,$
kg moles of N_2 per kg mole of injectant	$= \dfrac{\gamma}{100},$
kg moles of O per kg mole of injectant	$= 2 \cdot \dfrac{\epsilon}{100} = z,$

where x, y and z are defined in Section 11.1 and used in equations (11.1), (11.8) and (11.12). It can be seen from the above relationships that gas composition is readily represented in the stoichiometric operating equa-

Representing Complex Tuyère Injectants

tions using the x, y, z variables already discussed in the main body of Chapter 11. It must be remembered, however, that the quantity of gaseous injectant must be expressed in terms of kg moles of gas where 1 kg mole is equivalent to 22.4 Nm^3.

(b) Heat demand effects (Section 11.4)

The wustite reduction zone heat demand for an injectant (per kg mole of injectant) is given by

$$D^I = -H^f_{298_{\text{Injectant}}} + x \cdot [H^\circ_{1200} - H^\circ_{298}]_C + y \cdot [H^\circ_{1200} - H^\circ_{298}]_{H_2}$$

$$+ z \cdot \tfrac{1}{2} \cdot [H^\circ_{1200} - H^\circ_{298}]_{O_2} + \Sigma_{nj} \cdot [H^\circ_{1200} - H^\circ_{298}]_j,$$

where j represents components other than C, H and O (e.g. N_2) and n_j is moles of these components per mole of injectant.

The term $-H^f_{298_{\text{injectant}}}$ is, it will be remembered, the energy required to dissociate the injectant into C, H_2 and O_2 (and N_2). For a complex injectant, this term will not be known. It is, however, related to the heat of combustion H^C_V of the injectant (*Handbook of Chemistry and Physics 1977*, p. D279) by:

$$22.4 \cdot H^C_V = x \cdot H^f_{298_{CO_2}} + y \cdot H^f_{298_{H_2O_{(l)}}} - H^f_{298_{\text{injectant}}}$$

where H^C_V is the gross heat of combustion (298 K; to liquid H_2O; a negative number) kJ per Nm^3 of gaseous injectant; and x, y and z are as discussed above.

In final equation form, then,

$$D^I = (22.4 \cdot H^C_V - x \cdot H^f_{298_{CO_2}} - y \cdot H^f_{298_{H_2O_{(l)}}})$$

$$+ x \cdot [H^\circ_{1200} - H^\circ_{298}]_C + y \cdot [H^\circ_{1200} - H^\circ_{298}]_{H_2} + z \cdot \tfrac{1}{2} \cdot [H^\circ_{1200} - H^\circ_{298}]_{O_2}$$

$$+ \frac{\gamma}{100} \cdot [H^\circ_{1200} - H^\circ_{298}]_{N_2}$$

(kJ per kg mole of injectant).

AII.2 Injectants with Known Weight Percentages of Carbon and Hydrogen and Known Heats of Combustion

Injectants such as coal and oil are best treated on a weight basis from both stoichiometric and heat demand points of view.

(a) Stoichiometric effects

Consider 1 kg of oil with composition:

θ wt.% C,
λ wt.% H,
ϕ wt.% N,
ψ wt.% O.

Each kg of this oil contains:

	kg	kg mole
C	$\dfrac{\theta}{100}$	$\dfrac{1}{12} \cdot \dfrac{\theta}{100}$
H_2	$\dfrac{\lambda}{100}$	$\dfrac{1}{2} \cdot \dfrac{\lambda}{100}$
N_2	$\dfrac{\phi}{100}$	$\dfrac{1}{28} \cdot \dfrac{\phi}{100}$
O	$\dfrac{\psi}{100}$	$\dfrac{1}{16} \cdot \dfrac{\psi}{100}$

These terms may be represented in the stoichiometric equations by

$$n_C^A = n_C^{coke} + \frac{1}{12} \cdot \frac{\theta}{100} \cdot W^I - (C/Fe)^m, \tag{11.1A}$$

$$n_O^B + \frac{1}{16} \cdot \frac{\psi}{100} \cdot W^I + 1.06 = 1.3 \cdot n_C^A + 0.38 \cdot \frac{1}{2} \cdot \frac{\lambda}{100} \cdot W^I, \tag{11.8A}$$

where W^I is the weight (kg) of injectant per kg mole of product Fe.

Representing Complex Tuyère Injectants 215

(b) Heat-demand effects

In the case of solid and liquid injectants the heat demand of the injectant is best expressed by

$$D_W^I \text{ kJ per kg of injectant.}$$

Furthermore, the heat of dissociating solid or liquid injectants into C, H_2, O_2 and other components is related to the heat of combustion of the injectant by

$$H_W^C = \frac{1}{12} \cdot \frac{\theta}{100} \cdot H_{298\ CO_2}^f + \frac{1}{2} \cdot \frac{\lambda}{100} \cdot H_{298\ H_2O_{(l)}}^f - H_{W\ 298\ \text{injectant}}^f$$

where

H_W^C = heat of combustion of the injectant (298 K, to liquid H_2O and gaseous CO_2) a negative number;

$-H_{W\ 298\ \text{injectant}}^f$ = heat of dissociation of the injectant;

both kJ per kg of injectant.

In these terms, the wustite reduction zone heat demand of the injectant (per kg) is

$$D_W^I = \left\{ H_W^C - \frac{1}{12} \cdot \frac{\theta}{100} \cdot H_{298\ CO_2}^f - \frac{1}{2} \cdot \frac{\lambda}{100} \cdot H_{298\ H_2O_{(l)}}^f \right\}$$

$$+ \frac{1}{12} \cdot \frac{\theta}{100} \cdot [H_{1200}^\circ - H_{298}^\circ]_C + \frac{1}{2} \cdot \frac{\lambda}{100} \cdot [H_{1200}^\circ - H_{298}^\circ]_{H_2}$$

$$+ \left(\frac{1}{16} \cdot \frac{\psi}{100} \right) \cdot \frac{1}{2} \cdot [H_{1200}^\circ - H_{298}^\circ]_{O_2} + \frac{1}{28} \cdot \frac{\phi}{100} \cdot [H_{1200}^\circ - H_{298}^\circ]_{N_2}$$

and equation (11.12) becomes

$$D^{WTZ} + W^I \cdot D_W^I = n_C^A \cdot (198\,000) + \frac{1}{2} \cdot \frac{\lambda}{100} W^I \cdot (95\,000) + E^B \cdot n_O^B.$$

(11.12A)

APPENDIX III

Slag Heat Demands

This appendix describes methods by which the enthalpy effects of blast-furnace slag can be included in blast-furnace enthalpy balances. As will be seen, the enthalpy required to form liquid slag (1800 K) from its component oxides (1200 K) can be represented in terms of a discrete heat demand, D^{slag}, kJ per kg of slag. For simplicity, the discussion is restricted to CaO–SiO$_2$ slags only, but the method can be extended to include Al$_2$O$_3$, MgO, etc.

For slags formed from CaO and SiO$_2$ only, the additional terms in the Section 11.4 enthalpy balance are:

Input enthalpy* $\quad n_{SiO_2}^{slag} \cdot H^{\circ}_{1200} ; \quad n_{CaO}^{slag} \cdot H^{\circ}_{1200},$
$\qquad\qquad\qquad\qquad\qquad$ SiO$_2$ $\qquad\qquad\qquad$ CaO

Output enthalpy $\quad$ wt.$_{slag} \cdot H_{1800}$,
$\qquad\qquad\qquad\qquad\quad$ liquid slag
$\qquad\qquad\qquad\quad$ (kJ per kg of slag)

where $n_{SiO_2}^{slag}$, n_{CaO}^{slag} and wt.$_{slag}$ are kg moles of SiO$_2$ in slag; kg moles of CaO from limestone and burnt lime; and weight of slag, all per kg mole of product Fe. Of course, slag weight is given by:

$$\text{wt.}_{slag} = n_{SiO_2}^{slag} \cdot 60 + n_{CaO}^{slag} \cdot 56.$$

* $n_{SiO_2}^{slag} = n_{SiO_2}^{i} - (Si/Fe)^m,$
$n_{CaO}^{slag} = n_{CaO}^{i} + n_{CaCO_3}^{i}$

where $n_{SiO_2}^{i}, n_{CaO}^{i}, n_{CaCO_3}^{i}$ are kg moles of these components charged to the furnace per kg mole of product Fe.

Slag Heat Demands

Inserting these items into the enthalpy balance (and changing signs as described in Section 11.4) leads from equation (11.11) to

$$D^{wrz} + n^I \cdot D^I + \text{wt.}_{\text{slag}} \cdot H_{1800\ \text{slag}} - n^{\text{slag}}_{SiO_2} \cdot H^{\circ}_{1200\ SiO_2} - n^{\text{slag}}_{CaO} \cdot H^{\circ}_{1200\ CaO}$$

$$= E^B \cdot n^B_O - n^A_C \cdot (0.7 \cdot H^f_{1200\ CO} + 0.3 \cdot H^f_{1200\ CO_2}) - 0.38 \cdot y \cdot n^I \cdot H^f_{1200\ H_2O}.$$

(AIII.1)

In consolidated form, the slag-related enthalpy terms may be represented by

$$\text{wt.}_{\text{slag}} \cdot D^{\text{slag}} = \text{wt.}_{\text{slag}} \cdot (H_{1800\ \text{slag}} - u \cdot H^{\circ}_{1200\ SiO_2} - v \cdot H^{\circ}_{1200\ CaO}) \quad (AIII.2)$$

where $u = \dfrac{n^{\text{slag}}_{SiO_2}}{\text{wt.}_{\text{slag}}}$; $v = \dfrac{n^{\text{slag}}_{CaO}}{\text{wt.}_{\text{slag}}}$ (kg moles per kg of slag)

and D^{slag} is the slag heat demand, kJ per kg of slag.

This demand can be put in a more convenient form by making the substitutions (per kg of slag):

$$H_{1800\ \text{slag}} = H_{1200\ \text{slag}} + [H_{1800} - H_{1200}]_{\text{slag}},$$

$$H_{1200\ \text{slag}} = H^f_{1200\ \text{slag}} + u \cdot H^{\circ}_{1200\ SiO_2} + v \cdot H^{\circ}_{1200\ CaO}$$

(u and v as in equation (AIII.2))

which lead from equation (AIII.2) to

$$D^{\text{slag}} = H^f_{1200\ \text{slag}} + [H_{1800} - H_{1200}]_{\text{slag}} \quad \text{kJ (kg of slag)}^{-1}. \quad (AIII.3)$$

This equation shows that D^{slag} is exactly the heat required to form 1 kg of liquid slag (1800 K) from solid CaO and solid SiO_2, both at 1200 K (the thermal reserve temperature).

Suggested D^{slag} Calculation

Consider, for simplicity, that blast-furnace slag is made up of CaO and SiO_2 only and that its CaO/SiO_2 weight ratio is 1.2 (Table 1.1). One kilogram of this slag (0.55 kg CaO, 0.45 kg SiO_2) contains:

$$0.0098 \text{ kg mole of CaO} = v,$$
$$0.0076 \text{ kg mole of } SiO_2 = u,$$

which form a slag of the approximate composition:

$$0.0022 \text{ kg mole of } 2CaO \cdot SiO_2,$$
$$0.0054 \text{ kg mole of } CaO \cdot SiO_2.$$

Heats of formation (1200 K) of $2CaO \cdot SiO_2$ and $CaO \cdot SiO_2$ are −124 000 and −91 000 kJ (kg mole of compound)$^{-1}$, respectively (Appendix V), so that the heat of formation of 1 kg of slag is given by

$$H^f_{1200} \atop slag = 0.0022 \cdot (-124\ 000) + 0.0054 \cdot (-91\ 000)$$

$$= -750 \quad \text{kJ (kg of slag)}^{-1}.$$

The second term in the slag heat demand (equation (AIII.3)) is calculated from the slag enthalpy data given by Foerster,* i.e.

$$[H_{1800} - H_{1200}]_{slag} = 1000 \text{ kJ (kg of slag)}^{-1}$$

and the total demand is

$$D^{slag} = -750 + 1000 = +250 \text{ kJ (kg of slag)}^{-1}. \quad \text{(AIII.4)}$$

The above calculations do not consider the effects of Al_2O_3 or MgO, but it is estimated that the demand expressed by equation (AIII.4) is accurate to ±20% when the weights of all slag components are included in wt.$_{slag}$.

---

*Foerster, E. F. and Weston, P. L. (1967) *Heat Content of Some Blast-furnace and Synthetic Slags*, U.S. Bureau of Mines Report of Investigation 6886, Washington.

APPENDIX IV

Stoichiometric Data for Minerals and Compounds in Ironmaking

Compound	Mol. wt.	Wt.% M	Wt.% O	M/O weight ratio
Al_2O_3	102.0	53.0	47.0	1.13
CO	28.0	42.8	57.2	0.75
CO_2	44.0	27.2	72.8	0.37
$CaCO_3$	100.1	56.0 CaO	44.0 CO_2	
CaO	56.1	71.5	28.5	2.51
$CaO \cdot SiO_2$	116.2	48.3 CaO	51.7 SiO_2	
$(CaO)_2 \cdot SiO_2$	172.3	65.1 CaO	34.9 SiO_2	
Cr_2O_3	152.0	68.4	31.6	2.16
$Fe_{0.947}O$	68.9	76.8	23.2	3.30
Fe_3O_4	231.6	72.4	27.6	2.62
Fe_2O_3	159.7	70.0	30.0	2.33
H_2O	18.0	11.2	88.8	0.13
$MgCO_3$	84.3	47.9 MgO	52.1 CO_2	
MgO	40.3	60.3	39.7	1.52
MnO	70.9	77.4	22.6	3.43
Mn_3O_4	228.8	72.0	28.0	2.58
MnO_2	86.9	63.2	36.8	1.72
P_2O_5	142.0	43.6	56.4	0.78
SO_2	64.1	50.0	50.0	1.00
SiO_2	60.1	46.7	53.3	0.88

APPENDIX V

Enthalpies of Formation at Temperature T from Elements at Temperature T (H_T^f)

Values for compounds central to the analyses of this text (i.e. CO, CO_2, H_2O, $Fe_{0.947}O$) are accurate to within ±500 kJ. Sources are noted at the end of the table.

Units: kJ per kg mole of compound

Compound	298 K	1100 K	1200 K	1300 K
CH_4	−75 000	−91 000	−91 000	−92 000
C_2H_6	−84 000	−102 000	−101 000	−100 000
C_3H_8	−105 000	−129 000	−128 000	−127 000
CO	−111 000	−113 000	−113 000	−114 000
CO_2	−394 000	−395 000	−395 000	−395 000
$CaCO_3$ (from CaO, C and O_2)	−572 000	−563 000	−561 000	−559 000
$CaCO_3 \cdot MgCO_3$ (from carbonates)	−12 000	decomposes below 1100 K		
$(CaO)_3 \cdot P_2O_5$ (from CaO, O_2 and P_{2g})	−2 400 000	−2 364 000	−2 353 000	−2 341 000
$CaO \cdot SiO_2$ (from oxides)	−89 000	−91 000	−91 000	−91 000
$(CaO)_2 \cdot SiO_2$ (from oxides)	−126 000	−125 000	−124 000	−123 000

$Fe_{0.947}O$	−266 000	−265 000	−265 000
Fe_3O_4	−1 121 000	−1 096 000	−1 098 000	−1 095 000
Fe_2O_3	−826 000	−810 000	−811 000	−809 000
H_2O_l (ΔH_{298}^v vaporization = +44 000)	−286 000			
H_2O_g	−242 000	−248 000	−249 000	−250 000 (also −250 000 at 1500 K)
$MgCO_3$ (from MgO, C and O_2)	−511 000	decomposes below 1100 K		
MnO	−385 000	−386 000	−386 000	−386 000
Mn_3O_4	−1 387 000	−1 386 000	−1 386 000	−1 384 000
MnO_2	−521 000	−516 000	decomposes at 1120 K	
SO_2 (from gaseous S_2)	−362 000	−362 000	−362 000	−362 000
SiO_2	−911 000	−902 000	−901 000	−901 000

The data are from:

Stull, D. R., Prophet, H. *et al.* (1970) *JANAF Thermochemical Tables*, 2nd edition, United States Department of Commerce, Document NSRDS-NBS 37, Washington, June 1971,

with the exception of:

(a) CaO, $CaCO_3$, $CaCO_3 \cdot MgCO_3$, $(CaO)_3 \cdot P_2O_5$, $CaO \cdot SiO_2$, $(CaO)_2 \cdot SiO_2$, MgO, $MgCO_3$ which are from:
Robie, R. A. and Waldbaum, D. R. (1968) *Thermodynamic Properties of Minerals*, United States Geological Survey Bulletin 1259, Washington.

(b) C_2H_6, C_3H_8 which are from:
Scott, D. W. (1974) *Chemical Thermodynamic Properties of Hydrocarbons*, United States Bureau of Mines Bulletin 666, Washington.

APPENDIX VI

Enthalpy Increment Equations for Elements and Compounds, $[H_T^\circ - H_{298}^\circ]$

Enthalpy values calculated with these equations are within ±500 kJ of their published numerical equivalents. Sources are noted at the end of Appendix V.

Units: kJ per kg mole of element or compound

Substance	Temperature range, K				
	298–800 Top gas	800–1100	1100–1300 Wustite reduction	1100–1900 Bosh & Hearth	1800–2800 Flame temperature
Al_2O_3	$107T-32\,000$	$124T-46\,000$	$128T-50\,000$	$132T-55\,000$	—
C	$15.3T-4600$	$20.7T-8700$	$22.7T-11\,000$	$23.5T-11\,800$	—
CO	$30.2T-9100$	$32.5T-10\,800$	$34.2T-12\,600$	$35.3T-14\,000$	$36.6T-16\,400$
CO_2	$45.6T-14\,100$	$52.8T-19\,200$	$56.3T-23\,100$	$58.6T-26\,000$	$61.0T-30\,500$
CaO	$48.5T-14\,500$	$51.6T-17\,000$	$54.2T-19\,800$	$55.5T-21\,200$	—
$CaCO_3$	$106T-31\,500$	$123T-45\,100$	$129T-52\,200$	Decomposes above 1300 K	
$CaO \cdot SiO_2$	$108T-32\,300$	$122T-43\,600$	$128T-49\,500$	$128T-50\,000$	—
$(CaO)_2 \cdot SiO_2$	$158T-47\,000$	$186T-69\,000$	$189T-73\,000$	$218T-105\,000$	—
$(CaO)_3 \cdot P_2O_5$	$284T-84\,500$	$356T-142\,000$	$402T-193\,000$	$403T-194\,000$	—
Fe_S	$30.8T-9200$	$48.9T-23\,700$	—	$38.4T-11\,400$	—
$Fe_l(a)$ (M.P. ≈ 1800 K)	—	—	—	$44.0T-5800$	—

$Fe_{0.947}O$	$52.2T-15\,600$	$56.4T-18\,900$	$58.6T-21\,400$	$61.1T-24\,100$	—
Fe_2O_3	$135T-40\,200$	$158T-58\,800$	$142T-40\,600$	$144T-43\,000$	—
Fe_3O_4	$202T-60\,000$	$227T-82\,000$	$201T-53\,200$	$201T-53\,200$	—
H_2	$29.3T-8800$	$29.9T-9200$	$30.9T-10\,300$	$32.5T-12\,200$	$35.2T-17\,200$
$H_2O(g)$	$35.8T-10\,800$	$40.0T-13\,800$	$43.7T-17\,900$	$47.4T-22\,200$	—
MgO	$45.6T-13\,600$	$50.9T-17\,800$	$52.2T-19\,300$	$55.0T-22\,300$	—
$MgCO_3$	$102T-30\,500$	$126T-49\,500$	Decomposes below 1100 K		
$Mn_s(a)$	$31.0T-9200$	$44.3T-19\,900$	$38.2T-13\,200$	$48.6T-24\,600$	—
$Mn_l(a)$ (M.P. = 1517 K)	—	—	—	$48.6T-10\,000$	—
MnO	$49.2T-14\,700$	$53.7T-18\,300$	$56.9T-21\,800$	$57.1T-22\,000$	—
MnO_2	$68.2T-20\,300$	$77.3T-27\,600$			
Mn_3O_4	$167T-49\,600$	$182T-61\,900$	$199T-80\,200$	$238T-123\,000$	—
N_2	$30.0T-9000$	$32.1T-10\,500$	—	$34.4T-13\,000$	$36.4T-16\,500$
O_2	$31.6T-9600$	$34.3T-11\,500$	—	$36.2T-13\,500$	—
P_2g	$34.9T-10\,500$	$36.7T-11\,900$	$37.1T-12\,300$	$37.3T-12\,600$	—
S_2g	$35.2T-10\,600$	$36.9T-11\,900$	$37.2T-12\,300$	$37.4T-12\,500$	—
$Si_s(a)$	$23.6T-7000$	$26.4T-9300$	$27.2T-10\,200$	$27.1T-10\,100$	—
$Si_l(a)$ (M.P. = 1683 K)	—	—	—	$27.2T+40\,400$	—
SiO_2	$61.1T-18\,200$	$71.0T-25\,900$	$71.4T-24\,300^{(b)}$	$72.8T-26\,200^{(b)}$	—

(a) $[H^\circ_T - H^\circ_{298}]$
$M_l \quad M_s$

(b) $[H^\circ_T - H^\circ_{298}]$
cristobalite   quartz

APPENDIX VII

Numerical Values of E^B, Blast Enthalpy*

Temperature K	E^B (kJ per kg mole of O)
700	−39 000
800	−32 000
900	−24 000
1000	−16 000
1100	−8000
1200	0
1300	+8000
1400	+17 000
1500	+25 000
1600	+33 000
1700	+42 000

$$*E^B = \tfrac{1}{2} \cdot \left\{ [H^O_{T_B} - H^O_{1200}]O_2 + \frac{0.79}{0.21} \cdot [H^O_{T_B} - H^O_{1200}]N_2 \right\}$$

Answers to Numerical Problems

Answers within ±1% of those listed here indicate correct methods of calculation.

Chapter 1
1.1 (a) 1477 K; (b) 298 K; (c) 1773 K; (d) 1873 K; (e) 394 K
1.2 990 m^3
1.3 780 Nm^3
1.4 1.41 tonnes
1.5 (i) 452 kg per tonne of Fe; (ii) 2.10 kg moles per kg mole of Fe
1.6 53.3 Nm^3 or 76.1 kg

Chapter 2
2.1 1 Btu = 1.054 kJ
2.2 (a) 2.3 × 10^4 atmos.; (b) 34.6% CO, 0% CO$_2$, 65.3% N$_2$; (c) Virtually no effect
2.4 3.1 m sec^{-1}

Chapter 3
3.1 (a) 2450 K; (b) 2630 K; (c) 2400 K; (d) 2360 K; (e) 2140 K
3.2 (a) 65 kg per 1000 Nm^3 of air; (b) 0.105 kg moles of CH$_4$ per kg mole of O from dry air; 44 Nm^3 per 1000 Nm^3 of moist blast
3.4 (i) 5 × 10^{-3}; (ii) 9 × 10^{-6}

Chapter 4
4.1 (a) 1.5; (b) 1.26; (c) 0.24; (d) 2.13; (e) 1.89; (f) 1.89; (g) 1.46; (h) 1.02 CO; 0.87 CO$_2$
4.2 (a) 21% CO, 22% CO$_2$ 57% N$_2$; (b) 96 Nm^3 per kg mole of Fe; 1715 Nm^3 per tonne of Fe
4.3 (O/C)^g = 1.51
4.4 X_{CO} = 0.54; X_{CO_2} = 0.46
4.5 2.3 × 10^6 Nm^3
4.6 431 kg; 484 kg per tonne of Fe
4.7 +2·(Si/Fe)^m in left side

Chapter 5
5.1 (a) 486 000; (b) 494 000; (c) 521 000; (d) 543 000; (e) 556 000
5.2 525 000

Chapter 6
6.1 (a) 1935 Nm^3 per tonne of Fe; (b) 25% CO, 15% CO$_2$, 60% N$_2$
6.2 2.98 tonnes per minute
6.3 From 543 000 to 598 000 kJ per kg mole of Fe

Chapter 7
7.1 1360 Nm^3 per tonne of Fe
7.2 (a) None; (b) from 20% CO, 22% CO$_2$, 58% N$_2$ to 23% CO, 18% CO$_2$, 58% N$_2$
7.3 89%

Chapter 8
8.2 (a) 319 000; (b) 330 000; (c) 355 000; (d) 372 000; (e) 375 000

8.3 12 000 kJ per kg mole of O in dry blast air
8.4 −12 000 kJ per kg mole of O in dry blast air. The blast components must be heated to the thermal reserve temperature
8.5 It is always zero

Chapter 9
9.1 (a) 349 000; (b) 20 500; (c) 408 kg per tonne of Fe; (d) 1040 Nm³ per tonne of Fe; (e) 19% CO, 25% CO_2, 56% N_2
9.2 (a) 434 kg; (b) 1185 Nm³ of blast air, both per tonne of Fe
9.5 The O/Fe intercept at O/C = 0 is
$-[n_O^B + 2 \cdot (Si/Fe)^m + (Mn/Fe)^m]$

Chapter 10
10.1 400 K ± 20 (sensitive to 'rounding' errors)
10.2 25 K less than in Problem 10.1
10.3 (a) 1145 Nm³ per tonne of Fe; (b) 450 kg per tonne of Fe; (c) 1.22; (d) 13½%

Chapter 11
11.1 (a) 261 000 kJ per kg mole of C_3H_8; (b) 145 000 kJ per kg mole (6500 kJ per Nm³ of gas); (c) 4700 kJ per kg; (d) 2400 including sulphur, 2300 excluding sulphur, kJ per kg; (e) 68 000 kJ per kg mole (3000 kJ per Nm³) of gas; (f) −9100 kJ per kg mole of O_2; (g) 235 000 kJ per kg mole of H_2O
11.2 475 kg of coke per tonne of Fe
11.3 (a) 453 kg of coke; (b) 1280 Nm³ of moist blast per tonne of Fe
11.4 (a) 79 kg of oil per tonne of Fe; 80 kg per 1000 Nm³ of dry blast; (b) lowers carbon-from-coke requirement to 333 kg per tonne of Fe; lowers blast requirement to 983 Nm³ per tonne of Fe; (c) +3%
11.5 19½% CO, 22% CO_2, 3½% H_2, 5% H_2O, 50% N_2

Chapter 12
12.1 615 kg of coke; 1640 Nm³ of (moist) air; 240 kg of $CaCO_3$, all per tonne of Fe
2.2 555 kg of coke; 1430 Nm³ of (moist) air; 127 kg of CaO, all per tonne of Fe
12.3 600 kg of coke; 1585 Nm³ of (moist) air; 180 kg of $CaCO_3$, all per tonne of Fe; 3½% increase in production rate
12.4 540 kg of coke (12% decrease), 1475 Nm³ of (moist) air, 223 kg of $CaCO_3$, all per tonne of Fe; productivity increase of 11%
12.5 1 066 000 kJ/kg mole of dissolved P

Chapter 14
14.1 $42.1 per tonne of Fe
14.2 (a) 342 kg of coke (pure carbon), 171 Nm³ of CH_4, $49.7; (b) 455 of coke, 0 Nm³ of CH_4, $54.6; all per tonne of Fe
14.3 (a) $86.4; (b) $87.9; (c) $87.9; all per tonne of Fe
14.4 Increased to $87.1; 525 kg of coke; 26.8 kg of oil; 1450 Nm³ of moist blast; no oxygen; all per tonne of Fe
14.5 58.6 Nm³ of O_2; 15.0 kg of oil per tonne of Fe

List of Symbols

Appearing in Principal Equations

Symbol	Meaning	Units	Page
$(C/Fe)^m$	molecular ratio: C/Fe in product metal (assumed to be 0.25 (5 wt.! C) throughout the text)	$\left\{\dfrac{\text{kg moles of dissolved C}}{\text{kg mole of product Fe}}\right\}$	47
$\mathscr{D}$	whole furnace heat demand	kJ (kg mole of product Fe)$^{-1}$	60, 62
Wustite reduction zone heat demands			
D^C	for heating and dissolving carbon in product metal	kJ (kg mole of C in metal)$^{-1}$	161
D^{CaCO_3}	for decomposing charged $CaCO_3$ to CaO, C and O_2 at the thermal reserve temperature (extra heat demand due to charging $CaCO_3$ rather than CaO)	kJ (kg mole of $CaCO_3$)$^{-1}$	163
D^{Fe}	for reducing $Fe_{0.947}O$ and for heating and melting Fe	kJ (kg mole of product Fe)$^{-1}$	161
D^I	for decomposing 1 mole of tuyere injectant at its injection temperature to form its component elements at the thermal reserve temperature	kJ (kg mole of injectant)$^{-1}$	132
D^I_W	as above for 1 kg of injectant	kJ (kg of injectant)$^{-1}$	215
D^{loss}	for convective and radiative heat losses, per kg mole of product Fe	kJ (kg mole of product Fe)$^{-1}$	157
D^{Mn}	for reducing MnO and for heating Mn and dissolving it in product metal	kJ (kg mole of Mn in metal)$^{-1}$	160
D^{Si}	for reducing SiO_2 and for heating Si and dissolving it in product metal	kJ (kg mole of Si in metal)$^{-1}$	159
D^{slag}	for heating slag oxides and forming 1 kg of liquid slag	kJ (kg of slag)$^{-1}$	162

Symbol	Meaning	Units	Page
D^{wrz}	sum of wustite reduction zone heat demands (excluding injectant demands), per kg mole of product Fe	kJ (kg mole of product Fe)$^{-1}$	87, 88, 170
E^B	enthalpy in blast air above that which it would have at the thermal reserve temperature	kJ (kg mole of O in dry blast air)$^{-1}$	87, 89
$(Mn/Fe)^m$	molecular ratio: Mn/Fe in product metal	$\left\{ \dfrac{\text{kg moles of dissolved Mn}}{\text{kg mole of product Fe}} \right\}$	153
n_C^A	carbon taking part in wustite reduction zone reduction reactions = carbon leaving wustite reduction zone in ascending gases	kg moles of C (kg mole of product Fe)$^{-1}$	49
n_C^{coke}	carbon entering furnace in coke	kg moles of C (kg mole of product Fe)$^{-1}$	50, 126
n_O^B	oxygen entering furnace in dry blast air, i.e. not including oxygen from humidity or tuyère injectants	kg moles of O (kg mole of product Fe)$^{-1}$	49, 85
n_{CaCO_3}	limestone charged to the furnace	kg moles of CaCO$_3$ (kg mole of product Fe)$^{-1}$	155
n^I	quantity of tuyère injectant	kg moles of injectant (kg mole of product Fe)$^{-1}$	125
$(O/C)^g$	molecular ratio: O/C in carbonaceous portion of top gas		47
$(O/C)^{gwrz}$	molecular ratio: O/C in carbonaceous portion of the gas in the chemical reserve		79, 81, 144

$(O/H_2)^g$	molecular ratio: O/H_2 in hydrogenous portion of top gas		144
$(O/H_2)^{gwrz}$	molecular ratio: O/H_2 in hydrogenous portion of the gas in the chemical reserve		144
$(O/Fe)^x$	molecular ratio: O/Fe in iron oxides charged to the furnace (=3/2 for Fe_2O_3; 4/3 for Fe_3O_4)	kg moles of O (kg mole of product Fe)$^{-1}$	47
$(O/Fe)^{x\cdot wrz}$	molecular ratio: O/Fe in iron oxide in the chemical reserve (=1.06 for $Fe_{0.947}O$)	kg moles of O (kg mole of product Fe)$^{-1}$	79, 80
$(Si/Fe)^m$	molecular ratio: Si/Fe in product metal	$\left(\dfrac{\text{kg moles of dissolved Si}}{\text{kg mole of product Fe}}\right)$	153
T_B, T_g	temperatures of blast and top gas	K	
wt_{slag}	weight of product slag per mole of product Fe	kg of slag (kg mole of product Fe)$^{-1}$	216
W^I	weight of tuyère injectant per mole of product Fe	kg of injectant (kg mole of product Fe)$^{-1}$	214
x,y,z	moles of C, H_2 and O per mole of tuyère injectant	kg moles (kg mole of injectant)$^{-1}$	124

Index

Accuracy of model
  predictions 115, 173–177
  compared with industrial practice
    173–177
Activation energy
  coke gasification 36, illus. 37
  iron oxide reduction 43
Active Carbon, n_C^A
  coke requirement, related to 50, 168
  definition 49, 50
  equations 168
  graphical representation 52–55,
    98–106, 138–140, 144–148, 172,
    illus. 53, 55, 97, 99, 118, 139, 147
Active coke zone 16–19, illus. 18, 20
  reactions 36, 41
Adiabatic flame temperature see Flame
  temperature, tuyère
Agglomeration of blast furnace dust 9
Air blast see Blast air
Alkalis (K_2O, Na_2O)
  behaviour in blast furnace 8
  control 8
  flame temperature effects 34
  problems due to 8, 34
  removal in slag 8
Alumina (Al_2O_3) in blast furnace slag 8
Approach to equilibrium
  chemical reserve 38–39, 81, 117
  coke percolators 19, 36
Armour, movable 3, 26
Artificial variable in linear
  programming 193–195
Ash behaviour 21
Assumptions of mathematical
  model 77, 80–81, 114–116
  carbon gasification absent in top
    segment 77
  chemical reserve compositions 80–81
  chemical reserve zone 77

continuity between segments 76
equilibrium in chemical reserve 81
higher oxides absent in bottom
  segment 77
iron content of slag ignored 46
iron not reduced in top segment 77
steady-state conditions 45
summary illus. 78
testing of 114–116
thermal reserve zone 81
validity 114–116
wustite, only oxide descending into
  bottom segment 77
Auxiliary fuels see Tuyère injectants

Banking 10
Basicity, slag 8
  CaO/SiO_2 ratio 5
Bells, blast furnace illus. 2, 3
  bell-less top 26, illus. 27
Blast air see also Blast air requirement;
  Blast enthalpy Blast, hot; Blast
  temperature
  additives to 7 see also Tuyère
    injectants
  function 32
  hot, first use 10
  industrial requirement 5
  injectants with see Tuyère injectants
  moisture 33–34 see also Moisture in
    blast
  removal by refrigeration 34
  oxygen enrichment 11, 34 see also
    Oxygen enrichment
  pressure 16
  rate 5, 11, 16
    iron production affected by 10
    maximum and minimum 10

requirement *see* Blast-air requirement
temperature *see* Blast temperature
velocity 17
Blast-air requirement
  blast enthalpy effect on 96, illus. 99
  blast temperature effect on 96, 177
  calculation of 93–95, 101–103, 133–136, 140–143, 201
  CH_4 effect on 142
  coal injection effect on 177
  graphical representation illus. 53, 55, 97, 99, 118, 139, 147
  heat demand effect on illus. 99
  hydrocarbon injection effect on 177
  industrial, per tonne of hot metal 5
  limestone effect on (compared to CaO, $MgCO_3$ and dolomite) 156
  metallized ore effect on 177
  minimization of 204
  Mn in iron effect on 178
  natural gas injection effect on 177
  non-attainment of equilibrium, effect on illus. 118
  oil injection effect on 177
  oxygen enrichment effect on 134–137, 177
  predicted compared to actual 175–177
  production rate (iron) affected by 138
  reformed gas injection effect on 177
  Si in iron effect on 178
  thermal reserve temperature effect on 120
  tuyère injectant effects on 177
Blast enthalpy, E^B
  blast air requirement affected by 96, illus 99
  blast temperature, relationship with 89, 224
  calculation 94
  carbon requirement affected by 96, illus 99
  definition 87
  graphical representation 105, illus. 99
  negative/positive 89
  numerical values 224
  point H affected by 105, illus. 99, 171

Blast furnace illus. 2, 3
  advantages and disadvantages 12–13
  alkali control in 8
  approach to equilibrium in 36–39
  automatic control and operation 9
  bells illus. 2, 3
  burden 11, 26
  byproducts 8
  campaign life 10
  chemical behaviour and reactions 9–10, 16–41
  coke consumption, per tonne of metal 5
  control 9
  cooling 9, 11
  costs 12, 13
  counter-current nature 31
  critical hearth temperature 33
  description of process 1–13
  desulphurization in 8
  dimensions 4, illus. 3
  divided schematically through chemical reserve zone 76
  enthalpy requirements 31, 32
  erratic operation, channeling, flooding and hanging 10, 34
  free energy considerations 36, 37
  fuels 7, 177 *see also* Tuyère injectants
  fusion zone 19–21, illus. 20
  gas *see* Top gas
  height 3, 116
  improvements 11–12, 26
  impurity control 8, 178–179
  input details 3–9
  internal structure illus. 20
  internal volume, definition and example 3
  mathematical description *see* Mathematical model, blast furnace
  number, worldwide 1
  operating details 4–5, 9–11
  operating equations *see* Operating equations
  operation 9–11
  optimization *see* Optimization, blast furnace
  principal objectives 1
  production details 4–5, 175, 177

Index

production rate *see* Production rate, blast furnace
productivity 4
products 7–9
　per tonne of iron 4–5
　quenched 16
　raw materials 3–7
　　per tonne of iron 4–5
　reactions *see* Reactions, blast furnace
　slag *see* Slag, blast furnace
　stability of operation 10
　temperature *see* Temperature
　thermal behaviour 31–41
　top pressure 12
　tuyères *see* Tuyères
　visualization as two reactors 76
　world production using 1
Blast-furnace stoves illus. 2
Blast, hot *see also* Blast air, Blast temperature, Blast enthalpy
　advantages and benefits 10, 11, 32, 34
　blast requirement affected by 177
　coke requirement affected by 177
　enthalpy of, E^B 87
　first use 10
　flame temperature affected by *see also* Tuyère flame temperature 33
　furnace stability due to 10
　graphical representation 100, illus. 99
　heating of 9, illus. 2
　hydrocarbon injection affected by 34, 207
　representation in operating equations
　　whole furnace 109–110
　　wustite reduction zone 92, 168–169
　temperature, industrial 5, 7, 177
Blast temperature
　blast enthalpy, function of 87, 224
　blast-air requirement affected by 96, 177
　coke requirement affected by 96, 177
　flame temperature affected by 33, 34
　hydrocarbon injection affected by 34, 200, 208
　industrial 4–5, 7, 177
　necessity of high 7
　predicted effects compared to practice 177
Blowing rate, blast 11
Bosh 18, illus. 3
　reactions in 18–19, 41
　temperatures in 17
Bottom segment *see* Wustite reduction zone
Boudouard reaction (coke gasification) 29
Burden illus. 20
　advantages of even distribution 11, 26
　charging of 9, 26, illus. 3, 27
　components 1–7
　evenly distributed 26
　benefits of 26
　industrial methods of 26, illus. 3, 27
　layered structure 26, illus. 20
　layer thickness 26, illus. 20
　methods of charging 26, illus. 3, 27
　per tonne of iron 4–5
　prefluxed 7
　prereduced 176–178
　productivity improvements due to sized 3–7
Byproducts, blast furnace 8

Calcium carbide, external desulphurizer 8
Calcium carbonate *see* Limestone
Calcium oxide, CaO
　concentration in blast furnace slag 8
　enthalpy effects of 161–162
　forms of addition to furnace 7
　heat demand less than $CaCO_3$ 164
　low melting-point slag due to 7
　MgO partially replaces 156–157
　purposes in blast furnace 8
　sintered, in 7
　sulphur removal by 7–8
Calcium phosphate, $(CaO)_3 \cdot P_2O_5$ 7, 18, 166
Calculations *see also* Calculations, graphical
　blast air and carbon requirements 93–95, 133–137, 140–143, 201–202

Index

blast enthalpy (E^B) effect on illus. 99
heat demand (D^{wrz}) effect on 101
with hydrocarbon injection 140–143
with oxygen enrichment 133–137
coke requirement 140–143, 201–202
D^{wrz} and E^B effects on n_O^B and n_C^A 101, illus. 99
flame temperature 205–211
heat demands 157–164
of tuyère injectants 134, 140, 213, 215
heat of formation from heat of combustion 211, 213, 215
hydrocarbon injection effects 140–143, 177
operating parameters 93–95, 133–137, 140–143, 201–202
optimization 189–203
oxygen enrichment effects 133–138, 177
slag heat demand 216–218
thermal reserve temperature effects 120
top gas composition 51
including hydrogen 148–149
top gas temperature 109–112
tuyère injectants, effects on blast air and carbon requirements 133–137, 140–143
whole furnace 109–112
Calculations, graphical
blast air and carbon requirements 101–103, illus. 97
with CH$_4$ injection 144–148, illus. 147
with general injectant 144–148, illus. 147
with oxygen enrichment 138–140, illus. 139
top gas composition (O/C)^g 52, illus. 53
optimization 182–189
Campaign life, blast furnace 10
CaO see Calcium oxide
CaO/SiO$_2$ ratio in blast furnace slag 5
Carbon, active (n_C^A)
coke requirement, related to 50, 168
definition 49
equations 168
graphical representation illus. 53, 55, 97, 99, 118, 139, 147
Carbon balance, whole furnace 48–50, illus. 46
dolomite, magnesite incorporation 156
Carbon balances, wustite reduction zone basic 78–80, illus. 78
equations from 168
injectant incorporation 126, 212, 214, illus. 125
limestone incorporation 154–155
Carbon, inactive (C/Fe)^m 50
Carbon in iron
absorbed in coke percolators 18
carbon balance inclusion 50
(C/Fe)^m, definition 50
enthalpy balance inclusion 89
heat demand D^C 161
industrial level 7
saturation concentration 7, 50
Carbonates 154–157, 162–164 see also Dolomite, Limestone, Magnesite
Carbon dioxide see CO$_2$
Carbon monoxide see CO
Carbon oxidation 19, 36–38 see also Coke gasification
above chemical reserve, effect on coke demand 116
rate 23, 36–38, illus. 38
at tuyères 18–19, illus. 19
Carbon requirement, blast furnace see also Coke requirement
blast temperature effect on 96, 105, illus. 99
calculation of 94
with CH$_4$ injection 140–143, 177
with O$_2$ enrichment 133–137
graphical calculation of 101
with CH$_4$ injection illus. 147
with O$_2$ enrichment 138, illus. 139
heat demand effect on 96, 104, illus. 99
non-attainment of equilibrium effect on 117, illus. 118
prediction compared to practice 175–177

Index

thermal reserve temperature effect on 117, 119–120
tuyère injectant effects on 177
CH_4 see Natural gas
Channeling 10, 12
  effect on carbon and blast air requirements 117, illus. 118
  minimized by oxygen injection 12, 138
  minimized by top pressure 12
  minimized by uniform burdening 11
Charge, blast furnace see Burden
Chemical reactions see Reactions, blast furnace
Chemical reserve zone 38–40, illus. 39, 77
  assumption of equilibrium in 81, illus. 78
  basis for mathematical model 77
  compositions
    O/C 81
    O/Fe 80
    O/H_2 129, 145
  conceptual division through 76–78, illus. 77, 78
  departure from equilibrium in effect on blast and coke requirements ,117, illus. 118
  furnace height effect 116
  iron in 115
  magnetite in 115
  stability of operation due to 40
  temperature 38, illus. 77
  top boundary 40
CO
  Boudouard reaction, production by 9, 19, 22, 36
  concentration profile in furnace 39
  concentration in top gas 5, 8, 174
    calculation 51, 94, 148
  equilibrium with $CO_2/Fe/Fe_{0.947}O$ 37
  excess above chemical reserve 24, 38–40, 115
CO/CO_2 ratio
  in chemical reserve zone 39
  equilibrium
    with carbon 19, illus. 37
    with iron oxides 23, 24, 25, illus. 37
  $Fe/3O4$ 114, illus. 37
  non equilibrium in chemical reserve 117
  related to O/C ratio 47
CO_2
  from carbonates, inclusion in equations 154–157
  concentration in chemical reserve 38–39
  concentration in top gas 5, 8, 174
    calculation 51, 94, 148
  concentration in tuyère raceways, illus. 19
Coal injection into blast furnace 7, 177
Coke see also Coke gasification, Coke requirement
  ash 21, 210
  carbon balance, representation in, n_C^{coke} 50, 126, 154, 168
  coals for 6
  composition 6, 177
  constraint, optimization 200
  cost 13, 197, 201
  flame temperature incorporation 206, illus. 33
  industrial requirement, per tonne of iron 5, 175
  layered in blast furnace illus. 20
  moisture content 6, 149
  minimization of requirements for 181–203
  minimum in blast furnace for charge support and gas flow 6, 200
  percolators 18–19, illus. 18
  price relative to hydrocarbons 11
  production, properties, purposes 6
  raceway behaviour 16–17
  reactivity; effect on thermal reserve temperature 117–120
  reactivity; effect on blast, coke requirements 117–120
  replaced by tuyère injectants 140–143, 177
  size 6
  slits 20–21, illus. 21
  strength 6
    required in fusion zone 21
  sulphur content 6
  thermal reserve temperature affected by 117–120

Index

Coke gasification (Boudouard reaction) 19
  above chemical reserve (none) 39
  approach to equilibrium 36–37
  ceases below 1200 K 23
  cooling effect 22
  kinetics 23, 36–38, illus. 38
  reverse reaction (sooting) 37
Coke requirement, blast furnace smelting *see also* Carbon requirement
  active carbon, related to 50, 168
  blast temperature effects 96, 177
  calculation 142, 201
  CH_4 injection effect on 140–143
  coal injection effect on 177
  heat demand effect on 96
  industrial 5, 175
  limestone effect on 156
  minimum for charge support and gas flow 200
  Mn in iron effect on 178
  moisture in blast effect on 199
  natural gas injection effect on 140–143, 177
  oil injection effect on 177
  oxygen enrichment effect on 133–137, 177
  prediction 167–170, 202–206, 232
  prediction compared to practice 175–177
  prereduced (metallized) ore effect on 176–177
  reformed gas injection effect on 177
  Si in iron effect on 178
  tuyère injectant effects on 123–150, 176–177
Combined stoichiometry/enthalpy equations
  whole furnace 66
  wustite reduction zone 92, 169
Combustion, heat of 210, 213, 215
Complex tuyère injectants 207, 212–215
  flame temperature incorporation 207
  operating equation incorporation 212–215
Composition profiles *see* Gas composition
Composition ratios 47

$(C/Fe)^m$ 47
$(Mn/Fe)^m$ 153
(O/C) 47, 144
(O/Fe) 47
(O/H_2) 144
$(Si/Fe)^m$ 153
Compositions
  coke 6
  gas *see* Gas composition, Top gas composition
  iron 7
  ore 3
  slag 8
  top gas *see* Top gas composition
Computer programme
  for comparing model predictions with practice 173–179
  optimization (linear programming) 191–195
  strategy 173
Constraints, optimization *see* Optimization, blast furnace
Continuity, adhered to by mathematical model 76, 81
Control of iron composition 1, 35, 178
Convective heat losses, blast furnace 157–159
  heat demand 159
Cooling of blast furnace 9, 11
Costs *see also* Prices
  blast furnace plant 12
  iron production 13
  minimum, determination by linear programming 181–203
Counter current process, advantages 31
Critical hearth temperature 33–35
Critical operating parameters 1
Cyclic reduction and gasification 21–23

D *see* Heat demand, wustite reduction zone
$\mathscr{D}$ *see* Heat demand, whole furnace
D^C 161
D^{CaCO_3} 162–164
D^{Fe} 161
D^I_{298} 132, 140, 213, 215
D^I_T 132, 134
D^{loss} 157–159
D^{Mn} 160

D^{Si} 159
D^{slag} 161–162, 216–218
D^{wrz} 87–88, 164, 170
Data, industrial blast furnace
  operations 4–5, 175, 177
Decomposition temperatures of
  carbonates 154, 156
Deflectors movable (movable
  armour) 26, illus. 3
Demand, heat see Heat demand
Design, blast furnace illus. 3
Desulphurization see Sulphur in iron
Dimensions, blast furnace 4, illus. 3
Direct reduction processes 176
Division (conceptual) of blast furnace
  through chemical reserve zone 76,
  illus. 77, 78
  advantage of 76
  enthalpy and stoichiometric conditions
    imposed by 77
Dolomite, $CaCO_3 \cdot MgCO_3$ see also
    MgO 7, 156
  decomposition temperature 156
  replacement for limestone 156
  limit 157
  sulphur removal by 156
Dust losses from blast furnace 9, 46
  decreased by uniform burdening,
    pellets and sinter 11
  extent 46
  recycle of 9, 46
  removal from blast furnace gas 9,
    illus. 2

Elemental balances 45–47
Endothermic reactions
  Boudouard reaction (coke
    gasification) 9, 29
  decomposition of limestone 163
  $Fe_3O/Fe_{0.947}O$ reduction by CO 10
  reduction of iron oxides 32
Enrichment, oxygen see Oxygen
    enrichment
Enthalpy see also Heat demand, Heat
    supply
  content
    calculated from heat of
      combustion 210
    related to heat of formation 60, 86

demand see Heat demand
data 220–224
deficit overcome 32
equations, whole furnace
  298 K inputs/outputs 58–63
  complete 109
equations, wustite reduction
  zone 168
  basic 88, 92
  injectants included 133
  summary 168
graphical representation illus. 99
high temperature 31
of formation see Enthalpy of
  formation
of mixing in Fe see Enthalpy of
  mixing
requirements of blast furnace 31–35
supply see Heat supply
Enthalpy (heat) balance, whole
  furnace 58–63, 109–110
  carbon in iron incorporation 110
  combined with mass balance 66
  demand/supply form 59–62
  advantages 60
  equation, complete 109–110
  equation, simplified 62
  hot blast incorporation 109–110
  top gas temperature
    inclusion 109–110
Enthalpy (heat) balance wustite
  reduction zone 84–89
  assumptions 84, illus. 77, 78
  carbon in iron incorporation 161
  demand/supply form 86, 92
  heat demands see Heat demands
  heat loss incorporation 157–158
  heat of mixing incorporation
    159–218
  hot blast incorporation 87
  limestone incorporation 162–164
  manganese in iron incorporation 160
  operating equation 92, 168
  phosphorus in iron incorporation 166
  silicon in iron incorporation 159–160
  slag incorporation 161–162,
    216–218
  summary of equations 168
  thermal reserve temperature
    assumption 84, illus. 77, 78

thermal reserve temperature
  effect 119
tuyère injectants
  incorporation 130–132, 213, 215
Enthalpy, blast see Blast enthalpy
Enthalpy content
  298 K 60
  calculation from heat of combustion 211
  data 220–223
  hydrocarbons 211
  temperature T 86
Enthalpy of formation
  calculation from heat of combustion 213, 215
  data 220–224
  definition 86
  slag, CaO·SiO$_2$, 2CaO·SiO$_2$ 216–218
  temperature T (assumption) 132
Enthalpy of mixing in iron
  carbon 161
  manganese 160
  phosphorus 166
  silicon 159
Equations, operating see Operating equations
Equilibria
  C/Fe/O 19, 23–24, 114, illus. 37
  H$_2$/Fe/O illus. 29
Equilibrium in chemical reserve zone
  approach to 38, 80–81, illus. 39
  departure from, effect on operating requirements 117, illus. 118
  departure from, physical meaning 117
  graphical representation illus. 39, 77
Equilibrium in coke percolators 36, illus. 19
Equilibrium diagrams
  C/Fe/O illus. 37
  H$_2$/Fe/O illus. 129
Errors in model predictions 173
Exit gas see Top gas
Exothermic reactions
  CO formation 32
  CO$_2$ formation 9
  iron-oxide reduction by CO 10

Fe balances
  whole furnace 48
  wustite reduction zone 78, 79
Fe/C/O equilibria see Equilibria
Fe/H$_2$/O equilibria see Equilibria
Fe metal absent in top segment 77, 115
Fe$_{0.947}$O (wustite)
  in chemical reserve, model based on 23–25, 77
  composition 21
  heat demand 161
  reduction by CO 21–23, 36–37, illus. 37
  reduction by H$_2$ illus. 129
  slag, none in 8
Fe$_2$O$_3$ see Hematite
Fe$_3$O$_4$ see Magnetite
Fines, pelletization and sintering of 3, 11
Flame temperature, tuyère 33–35
  adiabatic, definition 205
  alkali and SiO vaporization affected by 34
  blast moisture effect on 34, illus. 33
  blast temperature effect on 33
  calculation 205–211
  coke ash effect on 210
  complex hydrocarbon incorporation 207, 210–211
  constraint equations for optimization 200, 208–209
  control 33–34
  critical, low 33
  definition 205
  equations 208–209
  excessive, problems due to 34
  hearth temperature, relationship with 33
  humidity effect, illus. 33
    removal by refrigeration 34
  hydrocarbon injectants effect on 34, 208
  industrial practice 34
  lowering factors 34
  metal composition affected by 35, 178–197
  oil effect on 207
  operating difficulties from incorrect 34

Index

optimization constraint 200, 208–210
oxygen enrichment effect on 34
prediction compared to practice 175
premature slag formation due to excessive 34
raising factors 33–34
sulphur in fuel effect on 210
tuyère injectant effects 34, 207–211
Flooding of bosh 10, 138
Flow pattern of gas in blast furnace 20
velocities 25–26
Fluidization of charge 138
Fluxes *see also* individual flux
dolomite, $CaCO_3 \cdot MgCO_3$ 7, 156
lime, burnt, CaO 7, 156
limestone, $CaCO_3$ 7, 154–156, 162–164, 197
magnesite, $MgCO_3$ 156
purpose 7, 8
quantity per tonne of Fe 5, 163
self fluxing burden 7
Framework, model 65–73
Free energy considerations in blast furnace 36–37
Free energy of reaction
$CO + Fe_{0.947}O$ 23
$CO + Fe_3O_4$ 24
$CO + Fe_2O_3$ 24
$CO_2 + C$ 19
$H_2 + Fe_{0.947}O$ 129
Fuels
coal 6, 7, 177
coke *see* Coke
natural gas 7, 140–143, 177
oil 5, 7, 11, 177, 196, 197, 200, 207–211
reformed gas 177
replacement ratios 142, 177
top gas, for stoves 9
Fuel injection *see* Tuyère injectants
Fuel oil *see* Oil
Fuel requirements, prediction compared to practice 175, 177
Fusion of slag, premature, cause of hanging 34
Fusion zone 19–21, 41, illus. 21
coke ash in 21
coke strength, importance in 21
gas distribution through 21, illus. 20

importance of 20–21
layer compositions. 20
physico-chemical processes in 21, 41
temperature 17

Gangue minerals 3
blast requirement affected by 96
coke requirement affected by 96
Gas, blast furnace *see* Top gas
Gas channeling 10, 11, 117, 138
Gas cleaning 9, illus. 2
Gas composition *see also* Top gas composition
chemical reserve zone 38, illus. 39
coke percolator 36, illus. 19
profiles, blast furnace illus. 39
tuyère raceway illus. 19
Gas, dust removal from 9, illus. 2
Gas flow, in blast furnace illus. 20
Gasification of coke *see* Coke gasification
Gas, natural *see* Natural gas
Gas, reformed 177
Gas, residence time 26
Gas temperature *see also* Top gas temperature
chemical reserve zone 39, 81, 117–120
profiles in furnace illus. 17, 35
thermal reserve zone 35–36, 77, 117–120
Gas velocities
inside furnace 25–26
tuyère 16
Gas volumetric flow rate 11
Graphical calculations *see* Calculations, graphical
Graphical optimization procedures 184–189
Graphical representation *see also* Operating diagram
active carbon illus. 53, 55, 97, 99, 139, 147
blast enthalpy illus. 99
blast-air (O) requirements illus. 53, 55, 97, 99, 139, 147
carbon in iron 101–103
carbon requirement 101–103

Index

chemical reserve
  compositions 98–100, illus. 97
CH_4, tuyère injectant illus. 147
combined stoichiometry/enthalpy
  equation, whole furnace illus. 69
discussion 54
Equation 4 illus. 53, 55
Equation 7 illus. 97
Equation 9 illus. 97
Equation 9 plus tuyère
  injectants illus. 139, 147
H_2H_2O 144–146
heat demand (D^{wrz}) 100, illus. 99
iron-ore composition 55
limestone 171–172
manganese in iron 171–172
natural gas injection illus. 147
non-equilibrium in chemical reserve
  zone illus. 118
operating line see Operating line,
  Operating diagram
optimization procedures 184–189
oxygen removed from iron
  oxides illus. 55
oxygen supplied in air blast illus. 55
oxygen (pure), tuyère injectant illus.
  139
point H 100, 171–172
point W 100, 171–172
  operating line, never to right
    of 115
silicon in iron 171–172
stoichiometry/enthalpy
  equation illus. 97, 99
summary of intercepts, points,
  slopes 171–172
top gas composition illus. 53, 54
tuyère injectants 144–148, illus. 147
wedges, permissible operating illus.
  73
whole furnace equations 68–70, illus.
  69

H, point 100, 171–172, illus. 97
H_2 see Hydrogen
H_2/H_2O equilibria illus. 129
H_2/H_{20} ratio in chemical reserve 129
H_2O see Moisture
Hanging 34

Hearth illus. 3
  coke in 18
  critical temperature 33–35
  reactions in 18, 41
  temperature illus. 17, 19
    factors raising and lowering 33–34,
      178–179
Heat see Enthalpy
Heat balances see Enthalpy balances
Heat of combustion
  in injectant heat demand
    calculations 213–215
  in flame temperature
    calculations 211
Heat demand whole furnace $\mathscr{D}$ 59–62,
  110
  advantages of concept 60
  carbon in iron 110
  definition 60
  examples 62
  equation 60
  graphical representation 68, illus. 69
  industrial 110
Heat demands, wustite reduction zone
  (D) 86–89
  advantage of concept 86
  blast-air requirement affected
    by 105, illus. 99
  $CaCO_3$, D^{CaCO_3} 162–164
  carbon in iron, D^C 161
  CH_4 140
  carbon requirement affected by 105,
    illus. 99
  coke reactivity effect on 119–120
  complex tuyère injectants 212–215
  definition and equations 88
  dolomite, $CaCO_3 \cdot MgCO_3$ 156
  Fe production, D^{Fe} 161
  graphical representation 100, 104,
    illus. 99
  heat loss (convection plus radiation),
    D^{loss} 157–159
  hydrocarbons, D^I 132, 134, 140,
    215, 217
  industrial 164
  lime (burnt) 161–164
  limestone, D^{CaCO_3} 162–164
  magnesite, $MgCO_3$ 156
  manganese in iron, D^{Mn} 160
  minimum 113

natural gas 212–213
oil 214–215
oxygen 134
silicon in iron, D^{Si} 159
slag, D^{slag} 161–162, 216–218
summary 170
thermal reserve temperature effect on 119–120
total, D^{wrz} 170
tuyère injectants, D^I
  298 K 132, 140, 215, 217
  temperature T 132, 134
Heat exchange, counter current 31
Heat losses, blast furnace 157–159
  heat demand, D^{loss} 159
Heat of formation *see* Enthalpy of formation
Heat of mixing *see* Enthalpy of mixing
Heat supply, whole furnace (S) 60, 110
Heat supply, wustite reduction zone (S) 87
  independent of thermal reserve temperature 119
  Heat transfer, gas-solid 41
  counter current 31
Height of furnace 4, illus. 3
  evolution and importance 116
Hematite, (Fe_2O_3) 3
  reduction by CO 10, 25
  reduction by H_2 148
  reduction high in stack 25
  whole furnace heat demand 62
High top pressure 12, illus. 3
Higher oxide (Fe_2O_3, Fe_3O_4) reduction 23–25
  restriction to small depth 24
Hot blast *see* Blast, hot
Hot blast stoves illus. 2
  blast furnace gas, fuel for 9
Hot metal *see* Iron, blast furnace
Humidity in blast *see* Moisture in blast
Hydrocarbon tuyère injectants *see* Tuyère injectants
Hydrogen
  carbon replaced by 142
  equilibrium with iron oxides *see* Equilibria
  graphical representation 144–148, illus. 147

injectant component *see* Tuyère injectants
operating equations, inclusion in 128–133, 168–172, 212–215
production in coke percolator 34
reduction of $Fe_{0.947}O$ by 129
in top gas 5, 175

Improvements in blast furnace productivity 11–12
Impurities in blast furnace iron 7, 178
  blast and coke requirements affected by 178
  concentrations 7, 178
  equations incorporating 168–170
  graphical representation 171–172
  heat demands 159–161
  heats of mixing 159–161
  minimization of 178
  oxygen balance affected by 153–154
  reduction 18
Inactive carbon $(C/Fe)^m$ 50
Injectants, tuyère *see* Tuyère injectants, *also* the specific injectant
Internal volume, definition illus. 3
Iron balances
  whole furnace 48, illus. 46
  wustite reduction zone 78, 79, illus. 78
Iron, blast furnace
  carbon saturation 7, 50
  composition 7 *see also* Carbon in iron, Manganese in iron, Phosphorus in iron, Silicon in iron
  control 35, 178
  flame temperature effect on 35, 178
  recent trends 178–179
  cost 13
  desulphurization 7, 8, 179
  production rate 5
  production, world wide 1
  raw materials per tonne 4–5
  in slag, low level 8, 46
  steelmaking plant specifications 35, 179
  tapping frequency 9
  temperature 17
  use 1, 179

Iron in chemical reserve, effect on blast furnace operating requirements   115
Iron ore   6
composition   3
  Fe_2O_3 3
  Fe_3O_4 3, 57
  impurities in   3, 8
  metallized (prereduced)   176–178
  pellets   3, 11   see also Pellets, iron ore
  quantity per tonne of iron   4–5
  price   13, 197
  sinter   3, 11   see also Sinter, iron ore
  size   3, 6
Iron oxide reduction   see Reduction

Kinetics
  coke gasification (carbon oxidation)   23, 36, 117, illus. 38
  iron-ore reduction   43
  time for reaction   26

Layer thicknesses in blast furnace   26
Life, blast furnace   10
Limestone ($CaCO_3$) and its decomposition   154–156, 162–164
  carbon balance representation   155
  coke requirement affected by   156
  cost   13, 197
  decomposition   154, 163
  enthalpy balance representation   162–164
  equations incorporating   168–169
  flux, blast furnace   7   see also CaO
  fuel requirement affected by   156
  graphical representation   171–172
  heat demand D^{CaCO_3} 163
  industrial quantity per tonne of iron   163
  oxygen balance representation   155
  purpose in blast furnace   7, 8
  size   7
  stoichiometric equation representation   168–169
  sulphur removal by   7, 179
Linear programming   189–203   see also Optimization, blast furnace

Liquid iron   see Iron, blast furnace
Losses, dust   see Dust losses
Losses, heat   157–159

Magnesite   see Magnesium carbonate
Magnesium, desulphurizer   8
Magnesium carbonate (magnesite)   156 see also Dolomite Magnesium oxide
  carbon requirement not affected by   156
  decomposition temperature   156
  top gas composition affected by   156
Magnesium oxide
  advantage over CaO   156
  concentration in blast furnace slag   8, 156
  forms of addition   7
  slag viscosity effects   156
  desulphurizer, inefficient   156
  MgO/CaO slag ratio   156
Magnetite (Fe_3O_4)
  in wustite reduction zone effect on blast furnace operating requirements   115–116
  reduction by CO
    to $Fe_{0.947}O$ 23, 40
    to Fe   114–116
  reduction by H_2 152
  reduction high in stack   23
  whole furnace heat demand   62
Manganese in blast furnace iron   7, 153–154, 160, 178
  blast and coke requirements affected by   178
  concentration   7
  control   178
  enthalpy equation representation   160
  equations incorporating   168–170
  graphical representation   171–172
  heat demand, D^{Mn} 160
  heat of mixing   160
  oxygen balance representation   153–154
  recent trends   178
Manganese oxides
  form   154
  reduction of   19, 41, 153

Index

Mass balance equations *see* Stoichiometric Equations
Mass balances *see also* Carbon balance, Iron balance, Oxygen balance
  whole furnace 48–50, illus. 46
  wustite reduction zone 78–80, illus. 78
  including injectants 128–129, illus. 125
Mathematical model, blast furnace
  accuracy of predictions 175–177
  assumptions *see* Assumptions of mathematical model
  blast requirement prediction
    using *see* Blast requirement
  chemical reserve assumption 76–81, illus. 77, 78
  coke requirement prediction
    using *see* Coke requirement
  comparisons, prediction/practice 175–177
  composition ratios *see* Composition ratios
  discussion 92, 108–121
  enthalpy balance, whole furnace 109–110
    simplified 59–62
  enthalpy balance wustite reduction zone 84–90
  equations *see* Operating equations
  graphical representation *see* Graphical representation
  objectives 44
  optimization using 181–203
  predictions compared to practice 175–178
  replacement (coke) ratio predictions 140–143, 177
  steady-state assumption 45
  stoichiometric development, whole furnace 45–50
  stoichiometric development, wustite reduction zone 78–81
  summary of equations and graphical points 168–172
  thermal reserve assumption 81, 84, 117–120, illus. 77, 78
  tuyère injectants, inclusion in 123–150, 212–215
  uncertainty of predictions 117–120, 175–176
  validity 108–116
  whole furnace parameters, usefulness for prediction of 96
  whole furnace 50, 109
Maximization *see* Optimization
  productivity 137–138, 204
Metal *see* Iron, blast furnace
Metallized (pre-reduced) ore
  industrial and predicted effects on operating requirements 176–178
Midrex process 176
Millscale, source of Fe 6
Minimization *see also* Optimization
  blast air 137, 204
  cost 181–203
Model *see* Mathematical model
Moisture
  in blast air
    endothermic reaction with coke 34
    flame temperature affected by 33–35, 206–210, illus. 33
    heat demand 150
    industrial levels 5, 33
    $n^I_{H_2O}$ related to n^B_O 199
    refrigeration, removal by 34
  in burden
    affects top gas composition 149
    cools top gas 21
  in coke 6
  in top gas 8

Natural gas (tuyère injectant) 7, 177
  blast and coke requirements affected by 142, 177
  calculations (CH_4) 140–142
  flame temperature incorporation 210
  graphical calculations (CH_4) illus. 147
  heat demand 140, 213
  industrial quantities 177
  prediction/practice comparison 177
  price 203
  reformed 177
Nitrogen
  concentration in top gas 5
    calculation 51
  concentration in tuyère raceway illus. 19

oxygen enrichment effect 11, 34, 137–138
Non-attainment of equilibrium in chemical reserve zone 117, illus. 118

Objectives, blast furnace 1
  mathematical model 14
O/C ratio
  in chemical reserve, $(O/C)^{gwrz}$ 77, 81, 117
  CO/CO_2 relationship 47, 51
  definition 47, 144
  graphical representation 51–55
  in top gas, $(O/C)^g$ 48, 50
    calculation with hydrocarbon injection 148
O/Fe ratio
  in chemical reserve zone 77, 80
  definition 47, 48
  graphical representation 51–55
  ore 48
  wustite 80
O/H_2 ratio
  in chemical reserve, $(O/H_2)^{gwrz}$ 144–145
  definition 144
  graphical representation 144–146, illus. 147
  in top gas, $(O/H_2)^g$, calculation 148–149
Oil (tuyère injectant)
  advantages 11
  blast requirement affected by 177
  carbon balance representation 214
  coke replacement ratio 177
  enthalpy balance representation 215
  enthalpy content 211
  flame-temperature effect 207–211
    calculation 207
  heat of combustion 211
  heat demand, D^I_{oil} 215
  industrial quantities 5, 177
  maximum quantities 177, 196
  prediction/practice comparison 177
  price 197
  W^I_{oil} related to n^B_O 200
Operating data, blast furnace 4–5, 175–177

Operating details, blast furnace 9–12
Operating diagram illus. 53, 55, 97, 99, 118, 139, 147
  basic description 50–55, illus. 53, 55
  blast enthalpy
    representation 105–106, illus. 99
  blast requirement interpretation 101, 138, 146, illus. 99, 139, 147
  CH_4 representation illus. 147
  coke requirement interpretation 172
  discussion of physical meanings 54, illus. 55
  heat demand (D) representation 100, illus. 97, 99
  hydrogen representation 146, illus. 147
  injectants, tuyère;
    representation 146, illus. 147
  non-equilibrium illus. 118
  operating line illus. 53, 55
    blast enthalpy affects position of 105, illus. 99
    definition 52
    heat demand affects position of 104, illus. 46,
    intercepts, meaning of 172
    permissible position 115
    points on 100, 171
    slope, meaning 51–52, 172
    summary of intercepts, points and slopes 171–172
  oxygen enrichment
    representation illus. 139
  point H 100, 171
  point W 100, 171
  slopes, meaning of 51–52, 172
  top gas representation 51–53, illus. 53
  wedges of permissible operation 73
Operating equations 168–169
  4 49–50
  7 80–81
  8 88, 92
  9 92
  10 109
  graphical representation *see* Graphical representation
  heat demands in 153–164, 170
  implications of 96

Index

optimization using 181–203
predictions compared to
practice 173–178
summary 168–169
tuyère injectant incorporation 129, 131, 212–215
whole furnace 49–50, 109
Operating line *see* Operating diagram
Optimization, blast furnace 181–203
artificial variables 193–195
blast plus O_2 minimization 204
coke constraints 200
coke minimization 181
computer programme
(LNPRG) 192–195
constraints 186–188, 200
cost minimization 181–203
equation linearity required
for 181–182
example problems 182, 196
feasible solution 188
flame temperature constraints 186, 200, 208–210, illus. 187
equations 208–210
graphical
calculations 184–189 illus. 185, 187, 89
improving the optimum 194
infeasibility 188, 191
linear programming 181
objectives 181
objective function 191, 196
oil constraint 196, 200
operating equations as
constraints 186
oxygen constraint form 204
production rate maximization 204
physical constraints 186–189, 200, illus. 187, 189
simplex algorithm 191–195
slack variables 193–195
vertex 191
Ore *see* Iron ore
Oxygen balance, whole furnace 48, illus. 46
dolomite, magnesite
incorporation 156
equation 50
Oxygen balances wustite reduction zone
basic 78–79, illus. 78

equations resulting from 168
injectant incorporation 126–129, 214
limestone incorporation 155
manganese in iron
incorporation 153–154
silicon in iron incorporation 153–154
summary 168
Oxygen in blast, n_O^B *see* Blast air
requirement
Oxygen enrichment (O_2 tuyère injectant)
advantages and benefits 11, 133
balanced with hydrocarbon
injection 34
blast-air requirement affected
by 136–137
calculations 133–140
coke requirement affected
by 137, 177
flame temperature raised by 12, 34, 133, 208–210
calculation 206–207
furnace productivity affected by 11, 137–138
gas volumes lowered by 137
graphical calculation and
representation 138–140, illus. 139
heat demand $D_{O_2}^I$ 134
industrial levels 5, 177
maximum level 112
oxygen requirement (total) affected
by 137
price 197
production rate (iron) increased
by 11, 138
temperature 124, 134
Oxygen exchange, counter current 31
Oxygen requirement *see* Blast-air
requirement
Oxygen in top gas 47

Paul Wurth top 26, illus. 27
Pellets 3
composition 3
improvement of blast furnace
performance by use of 11
size 3
Penetration of blast 16
Percolators 18

Performance, blast furnace *see* Production rate
Phosphorus in iron 19, 41, 166
Pig iron *see* Iron, blast furnace
Points, H, W *see also* Operating diagram
  enthalpy effects on point H 104–106, illus. 99
  non-equilibrium (W) 117, illus. 118
  operating line never to right of W 115
  summary of coordinates 171–172
Predictions of mathematical model *see also* Blast-air requirement, Carbon requirement, Coke requirement
  accuracy 119–120, 173, 175
  applicable to whole furnace 76
  blast-temperature effects 96, 106, illus. 99
  compared to practice 173, 175
  dolomite effect (top gas only) 156
  heat demand effects 96, 105, illus. 99
  heat-loss effect 104
  limestone ($CaCO_3$) effect 156
  manganese effect 178
  non-equilibrium effects 117, illus. 118
  optimization usefulness 180
  oxygen enrichment effects 134, 137–138
  phosphorus effect 166
  productivity effects 138
  silicon effect 178
  slag fall effect 96
  thermal reserve temperature effects 117, 119–120
  top gas composition 51, 175
    including H_2/H_2O 148–149, 175
  tuyère injectant effects 134–138, 142, 176
  validity examined 108–116
Pre-fluxed sinter 7, 57, 164
Pre-reduced (metallized) ore 176–179
Pressurization of furnace
  industrial 5
  productivity increased by 12
Pressure, tuyère 16
Prices *see also* Costs
  blast furnace plant 12
  coke 13, 197

iron-ore 13, 197
limestone 13, 197
natural gas 203
oil 197
oxygen 197
Probes, blast furnace 35, 39
Products, blast furnace 8, 178–179
Production details, blast furnace 45
Production rate, blast furnace 10, 122, 138
  blast rate effect on 10, 117–118
  blast requirement effect on 122, 138
  blast temperature (enthalpy) effect on 96
  determining factor, oxygen input rate 10
  gangue effect on 96
  heat-demand effect on 96
  hydrocarbon injection, effect on 176
  improvements, reasons for 11
  industrial 4–5, 11
  maximum 10, 117, 138
  minimum 11
  oxygen enrichment effect on 11, 138
  sized burden improves 11
  slag effect on 96
  top pressure effect on 12
  tuyère injectant effects on 11, 138, 176
  wind rate effect on 10, 122

Q, convective and radiative heat losses 157–158
Quenched blast furnace information 16

Raceways, tuyère 16–19, illus. 18, 19
  gas composition 19
Radiative plus convective heat losses heat demand, D^{loss} 157–159
Rates
  blast-furnace iron production *see* Production rate, blast furnace
  coke gasification (carbon oxidation) 23, 36, 117, illus. 38
  reduction, iron oxide 43
Raw materials, blast furnace 3–7
Reactions, blast furnace 41 *see also* Reduction, Coke gasification

above fusion zone 21–23
above 1200 K isotherm 23–25
active coke zone 16–18, 41
 bosh 18–19
 cyclic 21–23
 fusion zone 21, 41
 hydrogen reduction 32, 126–129, 148
 slag formation 21
  premature 34
 summary 41
 top quarter of shaft 25
 tuyère raceway 18–19, illus. 19
 wustite reduction zone 79, 153–156
Reactivity, coke
 effect on predictions of model 117, 119–120
 effect on thermal reserve temperature 117
Reducing efficiency, blast furnace 8, 46
Reducing gases, generation of 9, 19, 41
 above fusion zone 21, illus. 22
 above 1200 K isotherm (none) 23
Reduction 41
 above 1200 K isotherm 23
 approach to equilibrium 36–40, 117, illus. 39, 118
 $CO/Fe_{0.947}O/CO_2/Fe$ 21–23, 36–41, 80–81
 $CO/Fe_3O_4/CO_2/Fe_{0.947}O$ 24
 $CO/Fe_3O_4/CO_2/Fe$ 115
 $CO/Fe_2O_3/CO_2/Fe_3O_4$ 24–25
 coke percolators, in 18
 cyclic with coke gasification 21, illus. 22
 equilibria illus. 37, 129 see also Equilibria
 free energies 23, 24, 129
 higher oxide 23–25
 hydrogen 32, 126–129, 148–149
 impurities (Mn, Si, P) 18, 41, 153–154
 rate, iron oxide 43
 top quarter of shaft 25
 wustite reduction zone 79, 153–156
Reformed natural gas 177
Replacement ratio 142
 coke by CH_4 142
 coke by hydrocarbons 177
Reserve zones 35–40 see also Chemical reserve zone, Thermal reserve zone
Residence times in furnace, gases and solids 25–26

Sensible heat, top gas 110
Shaft illus. 3
 reactions in 41, 114–116
Silica (SiO_2)
 coke impurity 6, 197
 ore impurity 3
 reduction of 19, 153, 179
  rate 179
 slag component 8, 216–218
Silicon in blast furnace iron 178–179
 blast-air requirement affected by 178
 blast temperature effect on 178
 coke/ore ratio effect on 178
 coke requirement affected by 178
 concentrations industrial 7
 control of 178–179
 cool furnace effect on 179
 enthalpy equation representation 159
 equations incorporating 154, 168–169
 graphical representation 171–172
 heat demand D^{Si} 159
 heat of mixing 159
 humidity (blast) effect on 178–179
 hydrocarbon injection effect on 179
 oxygen balance representation 154
 recent trends 178
Silicon monoxide 8, 34
Simplex algorithm 191
Sinter, iron ore 4–6
 advantages of 11, 164
 $CaCO_3$ heat requirement avoided by 164
 dust loss minimized by 11
 Fe_3O_4 in 57
 prefluxed 7, 164
 size 6
Size
 blast furnace 4, illus. 3
  disadvantage 12
 raw materials 3, 6–7
Slack variable (optimization) 193–195
Slag, blast furnace
 alkali removal by 8

Index

basicity ratio 8
CaO/SiO$_2$ ratio 5
coke ash inclusion delayed 21
composition, choice of 8
enthalpy balance
  incorporation 161–162
enthalpy content 218
hanging due to premature melting of 34
heat demand 161–162, 216–218
industrial quantity 5
iron composition control by 8, 178–179
iron-oxide content of 46
melting-point 33
MgO effects on viscosity and desulphurization 156
  maximum MgO/CaO ratio 157
premature melting of, problems caused by 34
purposes 8
silicon control due to 179
sulphur removal by 8
temperature 33, illus. 17
uses for 8
viscosity of 8
Slits, coke 21, illus. 20
Soot formation 37
Space velocity 25
Stability of blast furnace operation 10
Steady state, model assumption 45, 76
  wustite production and reduction 24
Steelmaking requirements 1, 35, 179
Steelmaking slag, source of Fe, CaO and MgO 7
Stelco-Lurgi process 176
Stoichiometric balances see Carbon balance, Iron balance, Oxygen balance
Stoichiometric data, compounds 219
Stoichiometric diagram see Operating diagram
Stoichiometric equation, whole furnace 44–50
  graphical representation 51–52, illus. 53, 55
Stoichiometric equations, wustite reduction zone 168
  combination with enthalpy equation 92
  development 75–81
  dolomite, no effect on 156
  limestone incorporation 154–155
  Mn incorporation 154
  oil incorporation 214
  oxygen enrichment
    incorporation 134
  Si incorporation 154
  summary 168
  tuyère injectants
    incorporation 128–129, 212, 214
Stoichiometric validity of mathematical model 113–114
Stoichiometry, blast furnace 24–25
Stoves, hot blast 124, illus. 2
Sulphur, coke impurity 6
Sulphur in iron 7
  CaO effect on 7
  control 7–8, 179
  MgO effect 156
  removal in slag 8, 179
  removal outside furnace (prior to steelmaking) 8
  source (coke) 6
Summary of operating equations and operating diagram details 168–172
Supply, heat
  in hot blast 87, 94
  whole furnace 60
  wustite reduction zone, S^{wrz} 87
Symbols in principal equations 228–229

Tapping of furnace 9
Temperature
  blast 5, 177 see also Blast temperature
  bosh illus. 17
  carbonate decomposition 154–156
  chemical reserve zone 23–24, 77, 81, 117–118
  critical hearth 33–35
  controls iron composition 35, 179
  cyclic reduction zone 23
  flame see Flame temperature
  injectant 124
  iron 17, 179
  isotherms in blast furnace illus. 17

Index

profiles in furnace illus. 35
  idealized 77
  slag 17
  thermal reserve see Thermal reserve temperature
  top gas 5, 175–176 see also Top gas temperature
Testing of mathematical model 108–116
Thermal balance see Enthalpy balance
Thermal behaviour, in blast furnace, summary of 41
Thermal efficiency, blast furnace 31
Thermal requirements see Heat demands
Thermal reserve temperature 23, 24, illus. 35, 77
  assumed (1200 K) 77, 81, 117–119, illus. 77
  blast requirement affected by 117, 119–120
  coke reactivity effect on 117, 119
  coke requirement affected by 117, 119–120
Thermal reserve zone 23–24, 35–36, 117, 119–120, illus. 35, 77
  causes 23–24
  importance in model 81, 117, 119–120
Thermal validity of model 108–113
Thermodynamic behaviour, blast furnace 36–40, 114, 117–120
Thermodynamic data 220–223 see also Enthalpy, Free Energy
Thermodynamic validity of model 114
Times available for reaction 26
Top gas see also Top gas composition, Top gas temperature
  dust loading 46
  dust removal from 9, illus. 2
  fuel value 9
  industrial composition 5, 8, 175
  quantity, per tonne of iron 6
  reducing capability 114
  use in hot blast stoves 9
Top gas composition 5, 8, 175
  calculation 51
    including H_2/H_2O 148–149
  equilibria, relationship to 25, 114–116
  graphical representation illus. 55
  hydrogen 5, 8, 175
    calculations including 148–149
    industrial 5, 8, 175
  $MgCO_3$ effect on 156
  moisture in burden effect on 149
  oxygen content 47
  predicted compared to practice 175
  tuyère injectant effects on 148–149
Top gas temperature 109–113
  calculation of 110–112, 121
  industrial 5, 175
  prediction/practice discrepancy 173, 176
Top pressure 12
  industrial 5
  productivity improvement due to 12
Top segment stoichiometric equation (including H_2) 148
Tuyères
  behaviour in front of 16–19, illus. 18
  gas compositions in front of illus. 19
  gas flow through, velocity 16
  number of 16
  pressure 16
  raw materials introduced through 7, 124, 176–177
Tuyère injectants 123–150 see also Moisture in blast, Natural gas, Oil, Oxygen enrichment
  advantages 11–12, 34–35, 176
  amounts used 5, 177
  blast-air requirement, affected by 135–137, 142, 177
  calculations incorporating 133–150
  carbon balance, inclusion in 126, 212, 214
  carbon requirement affected by 135–137, 142, 177
  CH_4 142
  coal 7, 177
  coke requirement affected by 11, 135–137, 142, 177
  complex, incorporation in equations 212–215
  cost of iron, affected by 11, 202
  enthalpy balance, inclusion in 130–132, 213–215
  injectant heat demand definition 132

flame temperature affected
by 34–35, 200, 207–210
equations 208–210
general 124, 144–146, 212–215
graphical calculations
CH_4 147
O_2 138–140
graphical representation 171–172, illus. 147
H_2O *see* Moisture in blast air
heat demands, D^I (298 K) 132, 140, 213, 215
heat demands, D^I_T (T K) 132, 134, 213, 215
hydrocarbons
blast-air requirement affected by 142, 176–177
calculations including 140–143
carbon requirement affected by 142, 176–177
coke requirement affected by 11, 142, 177
cooling due to 34, 179, 208
flame temperature affected by 34, 208
graphical calculation (CH_4) illus. 147
industrial quantities 5, 177
prices 197, 203
price effects 11, 202
production rate effects 176
purpose of addition 11, 34, 124, 176
replacement of coke, ratio 142–175
top gas composition including 148–149
hydrocarbons plus oxygen 34, 150
mass balances inclusion in 126–129
moisture *see* Moisture in blast air
natural gas *see* Natural gas
oil *see* Oil
optimization incorporating 181–202
oxygen *see* Oxygen enrichment
oxygen balance inclusion in 126–129
prediction compared to practice 177
production rate effects 137–138, 176
purpose of addition 11, 34, 124
reformed natural gas 177
tar 7

temperature of addition 124
top gas composition
including 148–149
two or more 150
Tuyère raceways *see* Raceways, tuyère
Tuyère zone 16–18, illus. 18, 19
reactions in 19, 41
Two reactors, blast furnace as 76–78, illus. 77, 78

Use of mathematical model for optimization 181–182

Validity of mathematical model tested 108–116
prediction compared to practice 175
Vaporization of water, cooling of top gas by 121
Velocity, blast through tuyères 16
Velocity, gas in furnace 25–26
channeling, flooding and fluidization due to excessive 10, 138
maximum 10, 117
oxygen enrichment, effect on 12, 137–138
sized burden, equalizing effect 11
top pressure effect on 12
Velocity, solids descending in furnace 26, 34
Vertex 191
Viscosity of slag 8
MgO effect on 156
Volume, blast furnace 4, illus. 3

W, point *see* Point W
Water *see* Moisture
Wedges, operating 72, illus. 73
Whole furnace equations
4 50
10 109
top gas composition, prediction by 51 with H_2 148–149
top gas temperature, prediction by 110–112
Wind *see* Blast
Working volume
definition 3
industrial furnaces 4

Wurth top 26, illus. 27
Wustite ($Fe_{0.947}O$)
　composition 21
　heat demand, D^{Fe} 161
　steady-state production/reduction creates chemical reserve zone 24
Wustite reduction by CO
　enthalpy of reaction 22
　equilibrium illus. 37
Wustite reduction by H_2 129
Wustite reduction zone
　carbon balances in 78, 168
　　including $CaCO_3$ 154–155
　　including injectants 126, 212, 214
　definition 79, illus. 78
　enthalpy balances in 58–61, 130–132, 157–164, 213–215
　equations from, summary 168–169
　Fe descent into 115
　Fe_3O_4 descent into 115
　effect on coke requirement 116
　gas compositions in illus. 39, 77
　non-equilibrium 117, illus. 118
　heat losses from 157–159
　limestone decomposition in 154–156, 162–163
　manganese reduction in 41, 153–154, 160
　$MgCO_3$, decomposes above 156
　operating equations from, summary 168–169
　oxygen balances in 79–81, 153–154
　　including $CaCO_3$ 155
　　including injectants 126–129, 212, 214
　phosphorus reduction in 41, 166
　reactions in 77, 153, illus. 125
　silicon reduction in 41, 153–154, 159
　temperatures in 35, 77

x (carbon in tuyère injectant) 124, 212

y (H_2 in tuyère injectant) 124, 212

z (oxygen in tuyère injectant) 124, 212
Zones, blast furnace 41 *see also* Chemical reserve zone Fusion zone, Thermal reserve zone